The Politics of
Technology
Assessment

The Politics of Technology Assessment

Institutions, Processes, and Policy Disputes

Edited by
David M. O'Brien
University of Virginia

Donald A. Marchand
University of South Carolina

LexingtonBooks
D.C. Heath and Company
Lexington, Massachusetts
Toronto

Library of Congress Cataloging in Publication Data

Main entry under title:
The politics of technology assessment.

  Bibliography: p.
  Includes index.
  1. Technology assessment. I. O'Brien, David M. II. Marchand, Donald A.
T174.5.P64   353.0085'6   81–47763
ISBN 0–669–04837–2   AACR2

Copyright © 1982 by D.C. Heath and Company

Published simultaneously in Canada

Printed in the United States of America

International Standard Book Number: 0–669–04837–2

Library of Congress Catalog Card Number: 81–47763

The open society, the unrestricted access to knowledge, the unplanned and uninhibited association of men for its furtherance—these are what may make a vast, complex, ever growing, ever changing, ever more specialized and expert technological world, nevertheless a world of human community.
—J. Robert Oppenheimer, *Science and the Common Understanding*

For
Joyce P. Marchand and Eva R. O'Brien

Contents

Contents

List of Tables
and Figure

Acknowledgments

This book introduces students to the politics of technology assessment. Some of the chapters were previously delivered as papers at a National Convention of the American Society for Public Administration and based on research funded by several foundations. The authors are grateful for the support received from the National Science Foundation, the Office of Technology Assessment, and the Council for Research in the Social Sciences. The editors also appreciate the Office of Technology Assessment and the National Commission on Air Quality for inclusion of portions of their reports. Finally, we are especially grateful for the commendable typing of Ms. Vivian E. Kaufman of the Russell Sage Foundation, New York, New York, and for the services of Ms. Doris Ballentine of the Bureau of Governmental Research and Service at the University of South Carolina.

Politics, Technology, and Technology Assessment

David M. O'Brien and
Donald A. Marchand

Politics and Technology: Ambivalent and Enigmatic Interaction

From a contemporary domestic or international perspective, politics and technology share more than ever an ambivalent relationship. In modern, postindustrial societies, scientific discovery and technological innovation form the very basis on which economies define productive processes and explore new markets.[1] Governments actively regulate and promote technological innovation through research-and-development (R&D) grants, subsidies, contracts, and experimental projects for which the private sector may not, at the moment, be persuaded of profitable investment. For lesser-developed nations, technological innovation is an essential component of socioeconomic transformation. Modern, postindustrial societies, however, are also challenged by the darker side of technological change. Politics has been inflated and infused with social concerns about potentially adverse and unanticipated consequences of advancing technologies and industrial processes. Indeed, in the last fifteen years, political and legal institutions—at home and abroad—have increasingly confronted vexatious disputes over nuclear technology, computers, and a telecommunications revolution, as well as developments in chemistry and biomedical technology.

The ambivalent, enigmatic relationship between politics and technology invites a taking of sides in public debates—both on the general question of the beneficial or pernicious effects of technological change and on the impact of particular technologies such as nuclear power and recombinant DNA (rDNA). Inexorably, divisions among the left, right, and center of the political spectrum are exacerbated as technologies assume larger, more-prominent roles in modern life.

Technology is a social force embracing a complex web of political, economic, cultural, and philosophical dimensions. The permeating presence and wide-ranging ramifications of advancing technologies induce polycentric, if not intractable, problems of individual cum collective assessment. Such controversy and conflict should not be surprising because, as Hans Morgenthau sagely reminds us, "Man is a political animal by nature; he is a scientist by chance or choice; he is a moralist because he is a man."[2] Philo-

1

sophical disagreements, intense value conflict, and political polarization arise inevitably (and especially acutely) in a liberal democratic polity; demands for assessing and controlling technology collide with the alleged self-regulation of technology by market mechanisms in the private sector. The disagreements, conflicts, and polarizations are further intensified by the irreducible frailties of the human intellect and scientific uncertainties about the precise contours and polycentric impact of applied technologies. Philosophical and scientific controversies, for instance, persist over basic interpretative models and methods of extrapolation from data for determining carcinogenic risk. Philosophical and normative conflicts perpetuate political polarization when the scientific basis for health, safety, and environmental regulation remains uncertain. They also induce—as Aaron Wildavsky and Ellen Tenenbaum document with respect to the national-energy-policy debate—a profound mistrust of the information adduced for regulation, thereby reinforcing rival policy conclusions.[3]

Technology, and hence technology assessment, remains inescapably problematic, given intense value disagreements, at times intractable scientific uncertainties, and competing political pressures. Nonetheless, three major yet divergent perspectives on the relationship between politics and technology have shaped public debates in the last fifteen to twenty years and continue to influence the politics of technology assessment.[4]

The Invisible Hand of Technology:
Technological Determination

A good deal of contemporary discussion about the relationship between politics and technology seems strongly reminiscent of Adam Smith's invoking of "the invisible hand" to describe the process of self-coordination and mutual adjustment whereby independent producers and consumers, guided by the market prices of goods, attain joint material satisfaction. "The technicist projection," to borrow Manfred Stanley's term, however, is grounded in an assumption of technological domination.[5] Technology appears autonomous, monolithic. Technology constitutes not merely a means to attain socioeconomic ends but rather an end in itself with socioeconomic consequences that are portrayed as inevitable and unassailable. Politics is ultimately determined by the forces of technological change; indeed, politics and socioeconomic relations are fundamentally dictated by technology.

Like Reinhold Niebuhr's "children of light" and "children of dark," the technicist projection embraces two very different attitudes toward politics and the process of technological change. Some commentators express profound pessimism about the prospects for individual freedom and public choice in a technological society. French philosopher and political sociologist, Jacques Ellul, for example, argues that technology is no longer a means by which a polity may achieve the good life.[6] Instead, technology poses its own imperatives, dominating and obfuscating traditional social values—a view frequently pressed in the early 1970s over rDNA research. Technological rationality informs and prescribes individual and collective decisions.[7] Herbert Marcuse, moreover, finds not only the emergence of tyranny of technocracy but also one to which most Americans are blissfully seduced.[8] Although the polity continues to exhibit considerable pluralism, pluralism remains ineffective in countering the more-fundamental forces of technology. Individuals, he argues, become one dimensional, manipulated by technocrats and their own false consciousness. The technological society is more than symbolic. The aggregate and cumulative impact of innumerable technological advances may lead, in Nobel laureate Joshua Lederberg's words, to a "technopathic" polity.[9]

By contrast, other commentators find technology to hold out a kind of utopian promise. Rather than a technological Leviathan, these commentators find salutory opportunities and prospects with the advent of a technological society. Indeed, they insist that a "technological fix" can be applied to many of the most pressing social problems.[10] This perspective undergirded the federal government's policies and programs promoting nuclear energy, at least until the mid-1970s. Technology, as it were, expands the horizon of human possibilities, forging new economic and political relations, and perforce enriching the life of the polity.

Despite the divergence of these two normative views of technological change, both share an important, basic assumption. Both presume, as Jack Douglas observes, "the necessity of some form of tyranny and the impossibility of human freedom—that is, both presuppose a kind of historical and technological determinism.[11] From either perspective, individual and collective choices in a technological society are dominated—indeed, dictated—by technocratic imperatives and an instrumental rationality. As Victor Ferkiss comments: "Politics and culture are relegated thereby to the role of purely dependent variables. In technological determinism not only is technology an unexcused, unwilled cause . . . , but technological change is the sole and irresistible cause of all changes in all other fields of human activity."[12] Political institutions and processes are thus shaped by the dynamic and transforming forces of technology. As such, the significance of politics withers. Politics, like ethics, religion, and culture, represents only another arena that gives expression to technological imperatives.

Technological Intervention and Political Mediation

An alternative perspective to that of technological determinism is the view expounded by Emmanuel Mesthene, director of the Harvard University

Program on Technology and Society. Undeniably, as Mesthene argues, technology influences political decision making in at least two ways: (1) Technology creates new opportunities and resources, yet it (2) also generates new obstacles and policy problems. Technology, for Mesthene, "has both positive and negative effects, and it usually has the two *at the same time and in virtue of one another.*"[13] The mixed blessings of technology, he claims, arise from the following series of events:

> (1) Technological change creates a new opportunity to achieve some desired goal; (2) this requires alternatives in social organization if advantage is to be taken of the new opportunity; (3) which means that the functions of existing social structures will be interfered with; (4) with the result that the other goals which were served by the older structures are now only inadequately achieved.[14]

Conflict and polarization develop when public and private organizations are not able to take advantage of new technological opportunities due to limitations of existing socioeconomic and political arrangements. In the late 1970s, for instance, nuclear technology was arguably thwarted by excessive governmental regulations, prolonged litigation, and changing economic conditions, in spite of the earlier technology push provided by the national government. Still, other problems arise when the negative side effects of new technologies are not controlled because existing political and legal institutions refuse or prove incapable of a timely response. Disputes over toxic substances, industrial pollution, and the disposal of hazardous wastes, for example, frequently bode such criticism of legislatures and the courts.

According to Mesthene, on the one hand, technology itself must be considered "value neutral." The opportunities and problems posed by technology, on the other hand, are political: "The strains that technology places on our values and beliefs, finally, are reflected in economic, political, and ideological conflict. That is, they raise questions about the proper goals of society and about the proper ways of pursuing these goals."[15] "In the end," he maintains, "the problems that technology poses (and the opportunities it offers) will be resolved (and realized) in the political arena."[16] In other words, the political arena appears to mediate the normative conflicts engendered by technology.

The degree to which politics may either exploit or limit technology nonetheless remains uncertain. Technology appears as a conditioning variable in a complex socioeconomic and political equation: "one among many factors exogenous in its origins to the political process having necessary, significant and definable effects on political ideas, institutions and actions."[17] Technology, furthermore, is more important than other historical and social factors due to its "monolithic" and "undirectional" nature.[18] For these reasons, Mesthene seems to hold that technology is, in

John McDermott's words, a largely "self-correcting system."[19] He is, however, also led to propose the urgent "need to develop new institutional forms and new mechanisms to replace established ones that can no longer deal effectively with the kinds of problems with which we are faced."[20] Mesthene, like Manfred Stanley, therefore concludes that the problem is "how to organize society so as to free the possibility of (human) choice and how to control our technology wisely in order to minimize its negative consequences."[21]

Thus, as with technological determinism, Mesthene postulates that politics is profoundly conditioned by the forces of technology. Unlike technological determinism, technology does not overshadow politics. Instead, technology merely intrudes upon or intervenes in the political arena. Politics mediates normative conflict and may either exploit or limit the effects of technology. Notably, technology is nonetheless presumptively monolithic, existing largely independently of social and political decisions.

The Interdependence of Politics and Technology

A third approach postulates the dyanmic interdependence and interchange between politics and technology. In contrast with the previous perspectives, politics and technology are not "like two ships colliding at sea."[22] Political institutions and processes structure and influence the development and diffusion of technology—noticeably, the technological push given to nuclear energy in the 1950s and 1960s and governmental funding for 90 percent of all biomedical research. Concomitantly, technology weighs on political institutions and processes. Most significantly, technology presents new issues of public choice and policy, which are exemplified by the debates over personal privacy in our information-oriented society that emerged in the 1960s with governmental proposals to computerize criminal-justice records and, again, in the 1970s over private-sector development of electronic funds-transfer (EFT) systems.

Science and technology, as Don Price perceptively remarked, "cannot exist on the basis of a treaty of strict nonaggression with the rest of society; from either side, there is no defensible frontier.[23] Philosophical disagreements, normative conflicts, and political polarizations are neither surprising nor infrequent, due to the polycentric impact of technology and because, in a democratic polity and pluralistic society, competing visions of the nature and means of attaining public goods and the public interest exist. Technology fosters political debate and polarization. Controversy often centers on the propriety and extent of governmental intervention. More specifically, during the 1970s there was considerable public debate over the legitimacy and prudence of direct regulation; for example, pollution standards for industries. No doubt, during the 1980s that debate will continue, but with greater attention to alternatives to direct governmental regulation. Alternatives to regulation, such as pollution rights, emission fees, and subsidies in controlling air pollution, remain no less controversial in promising effective economic incentives for channeling private interests toward the modification and utilization of technologies for public purposes.

A measure of the fascination with the politics of technology assessment therefore derives from the import of the philosophical and political differences over whether, when, and how the public or the private sector should respond to the development, diffusion, and consequences of technologies. These differences in turn are heightened by profound disagreements over what, when, and how technologies and their polycentric consequences should be assessed and evaluated. Thus, the politics of technology assessment reveals deep-rooted philosophical and normative disagreements and, in a democratic polity, leads to a dynamic, fragmented, and pluralistic set of political institutions and processes.

Technology Assessment and Politics: Pluralism, Fragmentation, Conflict, and Consensus

Significantly, and in contrast with evaluation research and other social-science-developed forms of policy analysis, technology assessment was introduced and promoted by policymakers and analysts within the national government. The idea of technology assessment originated with Congressman Emilio Daddario. In 1966, he proposed establishing a specific staff for Congress that would monitor technological developments and provide an early warning of their potential hazards and detrimental impacts.[24] A series of studies and congressional hearings subsequently led the National Science Foundation (NSF) to establish a technology-assessment program and, in 1972, Congress to create the Office of Technology Assessment (OTA). During the 1970s, various federal agencies also undertook technology-assessment projects. After almost a decade, the OTA remains the primary producer of technology assessments, although some studies continue to be funded by the NSF, federal agencies, and private industry.[25]

As initially proposed and developed, technology assessments constitute a narrow class of policy analysis. They are distinguished principally by the objective of identifying and assessing the secondary or tertiary consequences of technology. In the words of an early advocate and architect of technology assessments, Joseph Coates, policy analysts endeavor to "systematically examine the effects that may occur when a technology is introduced, extended, or modified, with special emphasis on those consequences that are unintended, indirect, or delayed."[26] Assessment of the secondary, rather than the primary (intended), effects of technology is crucial, Coates argues, because:

> In the long run, the unintended and indirect effects may be the most significant;
>
> Undesirable secondary consequences often are unnecessary and may be prevented by proper planning;
>
> First-order impacts usually are subject to extensive study in the planning stage.[27]

Technology assessment, in other words, aims at clarifying socioeconomic and environmental problems potentially attendant technological developments and thereby providing information to the public and the government —especially Congress—that will inform public-policy decision making as well as guide R&D planning and natural-resource allocation.

Technology assessments ostensibly identify, structure, and evaluate alternative policy and technological options. They are usually predicated on either of two analytical modes or approaches: (1) problem-oriented assessments and (2) technology-oriented assessments.[28] Problem-oriented assessments focus on the effects of a technology, like industrial air pollution, with the objective of clarifying and analyzing the damages, causal relationships, and technological cum policy alternatives. By comparison, technology-oriented assessments endeavor to identify the future, long-term consequences of new technologies—for example, lasers, digital computers, or space satellites.

Both kinds of technology assessment draw on interdisciplinary social- and natural-science research including quantitative and qualitative data. The scope and methods of technology assessments must necessarily be tailored to particular technologies and policy issues. In the last decade, the OTA, for example, conducted over 100 technology assessments on a broad array of subjects, ranging from traditional and alternative energy technologies, natural-resource allocations, and computers and telecommunications systems, to techniques for carcinogenic-risk assessment and advances in medical technology. More generally, the OTA, the NSF, and federal agencies completed technology assessments in five major areas: (1) monitoring the side effects of existing technologies and industrial process, (2) maintaining environmental quality, (3) recycling resources, (4) screening scientific R&D, and (5) evaluating new technological innovations.[29]

Technologies and technology-policy issues, of course, vary markedly in terms of their socioeconomic and environmental ramifications as well as their potential for generating philosophical and political disagreements. Measurement of social risks, costs, and benefits remains vexingly problematic because crucial value judgments must be made—for example, how much do we value human life and by what means do we determine the worth of a life? Moreover, there often exists only limited data and relative agreement on the models and methods for conducting cost-risk-benefit assessments. Technology assessments therefore must incorporate a variety of analytical and methodological approaches including cost-benefit and systems analysis, environmental-impact analysis, and carcinogenic-risk assessment.

Technology assessment nonetheless remains a distinctive category of public-policy analysis, whether characterized by its aim, scope, general methodology, or relation to decision making. Even though technology assessment may utilize cost-benefit analysis, for instance, as Joseph Coates points out, there are important differences between the two.[30] Whereas technology assessment purports to identify and evaluate the (potential and actual) impact of technological innovations, cost-benefit analysis assesses the profitability of projects and governmental programs based on a comparison of economic costs and returns. Furthermore, the former is designed for long-range planning and policy formulation, while the latter is mainly employed in the selection of specific alternative projects and immediate policy options. Likewise, environmental-impact analysis—as required by the National Environmental Policy Act—ostensibly constrains policy formulation and decision making, yet technology assessments may still draw on such analysis and data.[31]

The promise of, and prospects for, technology assessment's improving political decision making and policy formulation remains problematic. Advocates of technology assessment encounter the formidable objection that such studies are both futile and pretentious. The National Academy of Engineering in 1969, for example, concluded that it was virtually impossible to systematically analyze the polycentric ramifications of particular technologies.[32] At best, technology assessments clarify policy objectives and options with respect to specific technological issues such as alternative strategies for utilizing computer technology and achieving trade-offs between the informational needs of organizations and the privacy interests of individuals. Still, for those who find a potential for a "technopathic" polity, assessment of a limited number of technologies appears inadequate since "technopathies" stem from the cumulative interaction of a combination of technological advances.

A more-troubling and penetrating criticism attends to the inherent limitations of the human intellect and, hence, to the reliability of technology assessments as well. Unquestionably, humbling and vexatious are the frequent disagreements over analytical models and research methods, along with the limited data on which confidently to establish causal connections and to predict long range trends.[33] During the 1970s, dramatic reappraisals of the costs, risks, and benefits of nuclear and biomedical technologies occurred in part because of increased data, greater precision in research strategies, and refined assessment methodologies. From the perspective of the sociology of knowledge, the findings of social and natural science are inevitably relative and uneven.[34] Yet, the alternatives for legislators and other policymakers of simply relying on common sense or uninformed political preference are arguably less advantageous.[35] Moreover, undeniably the processes of conflict resolution and consensus building within political arenas are profoundly affected by the difficulties of achieving agreement within scientific communities. Thus, the net informational value of technology assessments will remain a source of continuing debate.

The role of technology assessment in politics remains no less controversial. One long-standing bugaboo stems from the dilemma and popularized drama posed by tensions between technocracy and democracy—whence the old adage, "Keep experts on tap, not on top." Technology assessment, like other proposals such as synoptic planning to systematically introduce rationality into political decision making, seems both illegitimate and ill fated in a system of free government and pluralistic society. However, proponents of technology assessment, unlike Platonic guardians, neither purport to identify the correct or single best course of governmental action or aim to preclude political debate, compromise, and decision making by elected representatives. To be sure, some advocates of technology assessment have taken umbrage at the reluctance of political institutions to respond to technological developments and to utilize technology assessments. Francois Hetman, for instance, observes:

> Institutions seem to have an in-built tendency to stick to their first or initial
> purposes and to erect a wall against changes and external influences. They
> end up by concentrating on the defense of their existence and perenniality
> instead of modifying their functioning or yielding to new forms of action.[36]

He continues by countering that "the existing policy-making process implies no guarantee that a balance would be achieved between the benefits to society and, (a) the economic and social costs to the various groups affected by the technology; (b) the costs of controlling the damage; and (c) the cost to various groups of limiting the use of technology.[37] Implicit in Hetman's observations is an acknowledgment that technology assessment does not replace politics or elevate experts above elected representatives in the process of governmental decision making. Indeed, Hetman and others agree that the production of information about the costs, risks, and benefits of new technologies has usually neither precipitated nor dictated governmental action. To the contrary, governments typically respond to technological developments only after deleterious consequences have proved demonstrable and/or public attention and pressure has become so focused and intense as to force technology-policy issues on political agendas.

At bottom, the claim of advocates of technology assessment is quite modest—namely, improving the quality and quantity of information about technology available to policymakers and the general public. Technology assessment ostensibly has merit in purportedly enlarging and enhancing the quality of the information base for individual, collective, and governmental decisions.

Crucial concerns (and issues requiring further research) nevertheless remain about the politics of technology assessment. A frequent problem with technology assessments is that policymakers' alleged need for information often outstrips scientists' ability to provide that information, or at least to provide it in a timely fashion.[38] There are, as previously noted, also issues concerning who should conduct, and by what means, technology assessments.

From a broad political and historical perspective, assessment of technological developments appears pluralistic, fragmented, and incremental, as well as characterized by considerable overlap, duplication, and redundancy. Within the national government, technological assessments are mainly conducted by the congressional OTA but also by the NSF, various federal agencies, and on occasion, by congressional and presidential study commissions. Assessment and control of technology, however, is in fact much more pluralistic, fragmented, incremental, and uncoordinated; particularly when the activities and interchanges between the public and private sectors are considered.

Basic scientific and technological R&D is primarily subject to review by scientific peer groups and under some federal-agency-sponsored programs. In addition, legislative- and executive-branch oversight may occur at the federal, state, and/or local levels of government when technological innovations receive special funding or pose alleged risks to public health and safety. The development and utilization of new technologies, furthermore, remains subject to various kinds of review by multiple institutions and organizations in both the public and private sectors. The uses and impact of technological and industrial processes are to some extent assessed and monitored by professional associations including industrial and trade associations and labor unions, special-interest groups, and insurance companies. Firms developing and manufacturing particular technologies perform a range of studies in response to market forces and regulatory standards. The marketplace and private-sector decision making is notably constrained by federal and state legislation, for example, on environmental and consumer protection, as well as being increasingly liable to federal and state judicial constructions of statutory and common-law provisions for patents, copyrights, trade secrets, and actions for public and private torts.

Assessment of technologies thus is actually extremely fragmented, inexorably incremental, largely uncoordinated, and contingent on the dynamic political pressures of multiple institutions and organizations in the public and private sectors. Hence, perhaps the most fundamental issues of the politics of technology assessment revolve around the management and dissemination of technical information to policymakers and the public.[39]

Collective and governmental responses to technology depend on minimizing conflict and reaching a reasonable consensus. Resolution of technology-policy disputes, like other public-policy disputes, is difficult within a pluralistic society and a system of free government. Yet, technology-policy disputes are especially vexatious due to the particular combination of cross-cutting normative concerns (for example, debates on the wisdom of governmental intervention and the use of regulatory standards or incentive schemes) and issues of causality (for example, disagreements over the costs-risks-benefits of particular technologies). Opposing normative views, imperfect information and scientific uncertainties about matters of causality intensify and prolong technology-policy disputes.

The politics of technology assessment, therefore, raises basic questions not only about who, when, and how technology and its polycentric ramifications should be assessed and controlled but also about whether and how technology assessments are communicated to the public and policymakers— whether increased information is indeed utilized and, if so, in what ways and with what result? Finally, do technology assessments merely exacerbate deep-rooted political conflicts or contribute to consensus building on technology-policy issues?

The Politics of Technology Assessment: Institutions, Processes, and Policy Disputes

This book introduces students to the politics of technology assessment in terms of a pluralistic, fragmented political system; a dynamic and diverse set of processes for technology-policy formulation; and the underlying normative conflicts and scientific uncertainties over issues of causality. The book combines a concern with explicating the role of different political institutions and policymaking processes with a consideration of specific technology-policy disputes. Part I examines the respective roles of the private sector, the Congress—and its OTA—and the executive branch, as well as the judiciary, in assessing and governing advanced technologies. Part II consists of several case studies, illuminating the complex and dynamic processes of assessing and/or resolving technology-policy disputes and illustrating the uses and potential contributions of technology assessment.

While the chapters do not address all of the questions posed by the politics of technology assessment, they do answer some and provide a background for further research inquiry. Moreover, our endeavor here is to highlight the major issues that demand a hard look at the costs and benefits of technology assessment. Technology assessment represents more than a narrow reform movement of the 1970s, which has largely run its course. It rather epitomizes a broader, more-philosophical approach to governance and pragmatic decision making—one that has great resilience and public acceptance. After a decade of technological near disasters like Three Mile Island (TMI) and disasters like Love Canal, in addition to continuing controversies over environmental pollution, nuclear- and chemical-wastes management, computerized data systems and telecommunications, the general public remains concerned about environmental quality and natural-resource allocation as well as skeptical of the promised benefits of technological innovation cum diffusion.[40]

Central to the politics of technology assessment is whether the government should undertake its own assessments of technologies or simply rely on the private sector and the competitive pressures of the marketplace. In chapter 2, Vary T. Coates and David M. O'Brien argue that public-sector assessments of technology are necessary due to several significant market imperfections, including the facts that firms usually consider only the first-order, or direct, impacts of technology and not their broader social and environmental consequences and that competitive pressures on firms discourage their taking into account the externalities, or social costs, of their activities. Moreover, based on a small survey of industries, the authors conclude that technology assessment, or more precisely, parallel activities, in the private sector should be clearly differentiated from, and not considered as substitutes for, public-sector technology assessments. Their preliminary findings have been supported by a more-recent and larger survey of 100 firms by Battelle Laboratories.[41] While Coates and O'Brien conclude that the public sector remains the primary area in which the social and environmental consequences of technologies are likely to be evaluated, they also suggest that technology assessment may prove useful in forging greater cooperation between the public and private sectors in, for example, environmental management. This view is further explored in chapters 5 and 9, which examine the role of the Environmental Protection Agency (EPA) and strategies for controlling air pollution.

Chapters 3, 4, and 5 critically examine the development and utilization of technology assessment in the national government and political process. Vary T. Coates in chapter 3 reviews, in historical perspective, the record of the NSF's technology-assessment program, the congressional OTA, and the conduct of technology assessment by federal agencies. After discussing the internal politics of, and external pressures on, these institutions and agencies, she concludes that, during the 1970s, technology assessments were not fully integrated into technology-policy processes. Moreover, in the 1980s it is doubtful that sufficient organizational and political incentives exist to further promote technology assessment as an integral component of congressional and administrative policymaking, even though it might prove useful in forging greater public- and private-sector cooperation in responding to the application of new technologies.

The political history and context of technology assessment at the OTA and in federal agencies, as Coates shows, has indeed been troubled. The OTA in particular, has had major difficulty in focusing its resources and attention on congressional concerns while at the same time preserving the integrity of technology-assessment activities. Although the OTA has the potential for assuming a unique role in legislative and policy processes, as David Whiteman documents in chapter 4, its role remains tenuous, dependent on congressional perceptions and support. Focusing on OTA reports on the coal-slurry pipeline and residential-energy conservation, he analyzes their strategic and substantive use in Congress and considers how the use, nonuse, and misuse of reports may affect the institutional prestige of the OTA.

The technology-assessment activities of federal agencies have been no less problematic. The agencies have tended to focus on immediate policy issues rather than to study the long-range social consequences of developing technologies. Moreover, apparently neither have assessments been integrated within agency policymaking processes, nor have agencies often coordinated their activities—except in the area of cancer-risk assessment. The technology-assessment activities of the EPA is given specific attention by Coates and in chapter 5 by Gregory A. Daneke and James L. Regens. Albeit technology assessment has played some role at EPA, Daneke and Regens argue that Congress enacted technology-forcing statutes that gave the agency missions that often proved both technologically and economically infeasible. They insist, however, that strategic planning and technology assessment at the EPA may prove essential to its survival and to designing incentive systems—instead of regulatory standards—for environmental protection as well as to enhancing public- and private-sector cooperation. The standard-setting process at the EPA and alternative economic approaches to air-pollution control are discussed further in chapter 9.

Within the framework of the Constitution and with the creation of the OTA, Congress has the principal authority and responsibility for assessing the impacts of technology and promoting the public health, safety, and general welfare. Prudently or not, Congress has greatly delegated its responsibilities and powers to federal agencies. However, even prior to regulatory agencies, courts historically adjudicated disputes arising from industrialization and technological developments. During the 1970s, judicial intervention was also invited by the expansion of congressional legislation and administrative regulations on health, safety, and environmental matters. Furthermore, recent critics of regulatory agencies advocate reliance on the marketplace and judicially enforced property rights for controlling the deleterious consequences of technological and industrial processes.[42] In chapter 6, David M. O'Brien examines developments in private law adjudication and the allure of a judicial/administrative partnership in regulatory politics and, specifically, in the resolution of science- and technology-policy disputes. He shows that the judiciary has indeed assumed important regulatory functions and a participatory role in the political dynamics of resolving complex, polycentric policy disputes. Judicial intervention has been due in part to the growing litigation of technological disputes. Litigation serves both as a means of obtaining compensation for particular damages and as a vehicle for direct citizen participation in the formulation of public policy. Nonetheless, he concludes that the limitations of judicial expertise, the institutional structure of the judicial forum, and the adjudicatory process dictate only a modest role for the courts in science- and technology-policy disputes. Congress and the executive branch, not the judiciary, remain the legitimate and primary institutions for settling the normative conflicts and issues of causality presented by technological developments.

The politics of technology assessment, as noted earlier, is not confined to the national-governmental process. Professional associations, special-interest groups, and state and local governments are also frequently drawn into technology-policy controversies. The role of these organizations and institutions is highlighted in Part II in the chapters dealing with the politics of biomedical and nuclear-energy technologies. In chapter 7, Bonita Wlodkowski places the rDNA debate in historical perspective. She shows how that debate was elevated during the 1970s from the scientific community to the policy agendas of state and local governments, Congress, and the federal bureaucracy. In chapter 10, Deborah D. Roberts also discusses in historical perspective the assessment of nuclear-energy technology but focuses on the intergovernmental network for nuclear-energy regulation and on the institutional changes since the TMI incident. Both chapters illustrate the diversity and dynamics of technology-policy disputes and the multiple institutions and organizations that exercise more or less influence, at different periods, in structuring technology-policy disputes.

The rDNA and nuclear-energy controversies also indicate how exceedingly difficult it is to acquire and to assess data on the costs, risks, and benefits of particular technologies and thereupon to achieve some consensus on public policies. Indeed, both chapters indicate that cost-risk-benefit assessments are likely to vary widely and to evolve with the course of particular technology-policy disputes. The problems of assessing the costs, risks, and benefits of technological and industrial processes are explored in greater depth in chapters 8 and 9. Chapter 8 contains excerpts from a report by the

OTA on the techniques for determining cancer risks from the environment. In addition to providing a useful discussion of the models and methods for identifying carcinogens and for extrapolating from data to estimates of human-cancer incidence, the chapter serves to illustrate the work of the OTA. The politics and problems of conducting cost-benefit analysis in health and environmental regulation are also examined in the following chapter. Chapter 9 contains portions of the National Commission on Air Quality study of the Clean Air Act and alternative approaches to air-pollution control, such as the EPA bubble policy, emission banking, and pollution fees.

Chapters 11 and 12 examine two computer-technology-policy controversies arising from our information-oriented society. In chapter 11, Steven W. Hays, Donald A. Marchand, and Mark E. Tompkins explore the problems of assessing the impact of computerized criminal-justice record systems and balancing organizational interests with individuals' privacy interests. In chapter 12, Kent W. Colton and Kenneth L. Kraemer turn to the difficulties in evaluating the impact of EFT systems. Both chapters draw attention to a curious but perhaps predictable feature of the politics of technology assessment—namely, that far stronger incentives and resources exist for pursuing technological innovation and implementation than incentives or resources for undertaking impact assessments. One result of this imbalance, the authors suggest, has been a failure to refine and utilize social- and natural-science techniques for technology assessment. Moreover, at times the lack of data and tested methods of assessment has led Congress and federal agencies to mandate regulatory standards that were neither technologically nor economically feasible. At other times, utilization of technology assessments has been undermined by inadequate identification, documentation, and quantification of the costs, risks, and benefits of particular technologies and alternative policy options. The authors remind us that much more is at stake than underfunding and inadequate research priorities, but instead fundamental issues of public choice and self-governance in a technology-oriented society.

Notes

1. Daniel Bell, *The Coming of Post-Industrial Society* (New York: Basic Books, 1973).

2. Hans Morgenthau, *Scientific Man versus Power Politics* (Chicago: University of Chicago Press, 1965), p. 168.

3. Aaron Wildavsky and Ellen Tenenbaum, *The Politics of Mistrust: Estimating American Oil and Gas Reserves* (Beverly Hills: Sage Publications, 1981).

4. This section is indebted to the work of Victor C. Ferkiss. For a further discussion, see Ferkiss, "Man's Tools and Man's Choices: The Confrontation of Technology and Political Science," *American Political Science Review* 67 (1973):973 and *Technological Man: The Myth and the Reality* (New York: New American Library, 1969).

5. See Emmanuel Mesthene's review of Manfred Stanley's analysis in *Harvard University Program in Technology and Society, 1964–1972: A Final Review* (Cambridge: Harvard University Press, 1972), pp. 225–231.

6. Jacques Ellul, *The Technological Society* (New York: Vintage Books, 1964).

7. For a further discussion, see Laurence Tribe, "Technology Assessment and the Fourth Discontinuity: The Limits of Instrumental Rationality," *Southern California Law Review* 46 (1973):617.

8. See Herbert Marcuse, *One-Dimensional Man* (Boston: Beacon Press, 1964).

9. Joshua Lederberg, "The Freedoms and the Control of Science: Notes from the Ivory Tower," *Southern California Law Review* 45 (1976):605.

10. See, for example, Alvin Weinberg, "Can Technology Replace Social Engineering?" in *Technology and Man's Future,* Albert Teich, ed. (New York: St. Martin's Press, 1972), pp. 27–34.

11. Jack Douglas, ed., *Freedom and Tyranny: Social Problems in a Technological Society* (New York: Alfred A. Knopf, 1970), p.13.

12. Ferkiss, "Man's Tools and Man's Choices," p. 974.

13. Emmanuel Mesthene, *Technological Change* (New York: New American Library, 1970), p. 26.

14. Emmanuel Mesthene, "The Role of Technology in Society, " in *Technology and Man's Future,* Albert Teich, ed. (New York: St. Martin's Press, 1972), pp. 130–131.

15. Mesthene, *Technological Change,* p. 148.

16. Mesthene, Ibid, p. viii.

17. Ferkiss, "Man's Tools and Man's Choices," p. 974.

18. Ibid.

19. John McDermott, "Technology: The Opiate of the Intellectuals," in *Technology and Man's Future,* Albert Teich, ed. (New York: St. Martin's Press, 1972), pp. 151–178.

20. Mesthene, "Role of Technology," p. 148.

21. Mesthene, *Technological Change,* p. 148.

22. Kenneth Laudon, *Computers and Bureaucratic Reform* (New York: John Wiley, 1974), p. 30.

23. Don Price, *Government and Science* (New York: New York University Press, 1954). See also Robert Merton, "The Normative Structure of Science," *Journal of Legal and Political Sociology* 1 (1942):115.

24. See, for example, Emilio Daddario, "Technology Assessment Legislation," *Harvard Journal on Legislation* 7 (1970):507.

25. For further discussion of the history and development of technology assessment, see Francois Hetman, *Society and the Assessment of Technology* (Paris: Organization for Economic Cooperation and Development, 1973), chapter 3 of this book; Derek Medford, *Environmental Harassment or Technology Assessment* (New York: Elsevier, 1973); Organization for Economic Cooperation and Development, *Methodological Guidelines for Social Assessment of Technology* (Paris, 1975); U.S. Congress, House, Committee on Science and Astronautics, Subcommittee on Science Research, *Technical Information for Congress,* 92nd Cong., 1st sess., 1971; and U.S. Congress, House, Committee on Science and Astronautics, *Technology Assessment: Processes of Assessment and Choice,* 91st Cong., 2d sess., 1969, Report of the National Academy of Sciences.

26. Joseph Coates, "Technology Assessment: The Benefits . . . the Costs . . . the Consequences," *The Futurist* (1971):225.

27. Ibid.

28. See U.S. Congress, *Technology Assessment,* and Hetman, *Society and Assessment of Technology,* pp. 61–63.

29. See, chapter 3 of this book and Hetman, *Society and Assessment of Technology,* pp. 263–330.

30. Joseph Coates, "What Is Technology Assessment?" *International Association for Impact Assessment Bulletin* 1 (1981):20.

31. National Environmental Policy Act of 1969, 42 U.S.C. §§4321–4361.

32. See U.S Congress, *Technology Assessment.*

33. For a further discussion, see chapter 9 of this book. See also Carol Weiss, *Evaluating Social Action Programs* (Boston: Allyn and Bacon, 1972).

34. See, generally, Robert K. Merton and Norman Storer, eds., *The Sociology of Science* (Chicago: University of Chicago Press, 1973); Michael Polanyi, *Personal Knowledge* (New York: Harper & Row, 1964); Thomas Kuhn, *The Structure of Scientific Revolutions* (Chicago: University of Chicago Press, 1962); and May Brodbeck, *Readings in the Philosophy of the Social Sciences,* 5th ed. (New York: Macmillan, 1970).

35. For a further discussion, see the debate in the following articles: David M. O'Brien, "The Seduction of the Judiciary: Social Science and the Courts," *Judicature* 64 (1980):8–21; Peter Sperlich, "Postrealism: Should Ignorance Be Elevated to a Principle of Adjudication?" *Judicature* 64 (1980):93–98; and O'Brien, "Of Judicial Myths, Motivations and Justifications: A Postscript in Social Science and the Law." *Judicature* 64 (1981):285–290.

36. Hetman, *Society and Assessment of Technology,* p. 333.

37. Ibid., p. 335.

38. For a further discussion of the use of social science by policy-makers, see Weiss, *Evaluating Social Action Programs.*

39. For a further discussion and survey of the literature, see Jack Knott and Aaron Wildavsky, "If Dissemination Is the Solution, What Is the Problem? *Knowledge: Creation, Diffusion, Utilization* 1 (1980):537–578.

40. See, for example, "Public Opinion on Environmental Issues," in *Environmental Quality,* Eleventh Annual Report of the Council on Environmental Quality, pp. 401–425 (Washington, D.C.: Government Printing Office, 1980).

41. See Kazuhiko Kawamura, "Technology Assessment as a Planning Tool," *International Association for Impact Assessment Bulletin* 1 (1981):25–31.

42. See, for example, Roger Meiners, "What to do about Hazardous Products," in *Instead of Regulation,* Robert Poole, Jr., ed. (Lexington, Mass.: Lexington Books, D.C. Heath and Company, 1981), p. 285.

Part I
Institutions and Processes

2

Technology Assessment and the Private Sector

Vary T. Coates and
David M. O.Brien

Technology, the Marketplace, and Politics

Technology assessment describes a form of public-policy analysis. Technology assessment was conceived as a means of improving public-sector decision making by identifying in advance the possible direct and indirect socioeconomic and environmental consequences of developing technologies. Contrary to some commentators, use of the term *technology assessment* to describe comparable analyses designed to assist private-sector decision making appears undesirable and counterproductive.[1] Applying the term to these activities obscures rather than illuminates. It is also misleading and deprecates some very useful and progressive developments in business management and planning.[2] Furthermore, insistence on technology assessment in the private sector tends to evoke resentment and resistance in industrial management. This resentment and resistance, in turn, often prevents the private sector from appreciating and making use of the findings of public-sector technology assessments as well as from politically supporting governmental initiatives in technology assessment.

Public policy, as broadly defined by Carl Friedrich, denotes "a proposed course of action of a person, group, or government[al] [agency] within a given government that contains obstacles and opportunities which the policy as proposed intends to utilize and overcome in an effort to reach a goal, or realize an objective, or achieve a purpose."[3] Technology assessment was conceived and has evolved over a decade-and-a-half to enhance the formulation and implementation of the objectives and aims of public policy. The concept of technology assessment in this sense was introduced in 1966 by Congressman Emilio Daddario, who proposed establishing a special staff to provide Congress with early warning of hazards or detrimental impacts that might result from developing technologies, especially those encouraged by federal policies or developed with public resources.[4] Technology-assessment activities are presently conducted by various federal agencies, the technology-assessment program of the NSF, and by the OTA.

The authors express their appreciation to Ms. Thecla Fabian for helping to conduct the research and interviews for this chapter. The chapter is based on a report submitted to the OTA, U.S. Congress.

Technology assessment, broadly speaking, entails interdisciplinary social-science analysis designed to identify and clarify the social costs and consequences of technological and industrial processes—for example, potential environmental-pollution problems, public-health risks, economic and institutional disruptions, and other social costs associated with technological advances in the private sector and with large-scale public works. Identification and measurement of social risks, costs, and benefits inevitably remains vexingly problematic and politically controversial. Technology assessment thus evolved as not one research algorithm or model but as a variety of analytical techniques that are tailored to specific policy issues.[5] Technology assessment, however, remains a fundamentally different activity from feasibility studies (can an industry make a technological process work effectively and efficiently?) and market studies (how should a company package and sell a product?) conducted in the private sector. Technology assessment, moreover, neither purports to identify the correct, or single best, course of governmental action nor precludes political debate, compromise, and decision making by elected representatives. Instead, it aims at clarifying the nature of perceived or actual social problems attending technological developments, thereby providing information to the public and the government in order to guide R&D planning and natural-resource allocation, as well as enhancing governmental—especially congressional—decision making.

Public-sector assessments of technological developments appear necessary due to certain market imperfections. In this regard, as Milton Katz explains, there are two broad dimensions of technology assessment:

> [First,] a systematic comparative appraisal of the first-order effects of technology (electric power [for example] . . .) in relation to the . . . discoverable and foreseeable side effects (air pollution . . . and possible radioactive hazards in the present context). . . . [Second,] a search through the full range of technological possibilities for the one best designed to achieve the desired first-order while . . . minimizing the undesirable side effects.[6]

The marketplace, as Congressman Daddario and others noted when endorsing the idea of technology assessment, "only consider[s] the first-order consequences, and not the broader social implications" of developing technologies and industrial processes.[7] The immediate social consequences and costs of economic transactions, moreover, are normally not reflected in the price of goods, in part because of market imperfections—for example, the lack of information about the societal and environmental effects of industrial processes and consumer goods. Competitive pressures for investments and innovations in the market also discourage firms from taking into account the externalities of their activities—that is, the secondary effects of, and impact on, parties not directly involved in market transactions. The market discourages a firm from assessing and assuming (internalizing) the social costs of hazardous production processes or consumer goods when its competitors will not voluntarily do likewise. Thus social costs, or negative externalities like air pollution, accompanying industrial processes and consumer goods are not reflected in the price of a good determined solely by market forces.[8]

The import of technology assessment in the public sector is evident if we consider not only the dynamics of the marketplace but also the issue of corporate responsibility. It remains debatable whether the private sector has a responsibility to undertake assessments and to make subtle determinations as to what constitutes the public interest. Some technological risks may be socially acceptable, whereas others may arguably fall outside the bounds of public morality. One might argue, for instance, that introduction of a new pollutant, toxin, or carcinogen that poses a substantial health risk (whether or not that risk is covered by existing legal prohibitions and regulations) is unequivocally wrong and violates the social responsibility of industry. However, agreement on what constitutes unacceptable risk and on the imperatives of a public morality remains difficult to achieve in a heterogeneous society and pluralistic political process. The nature of private enterprise and the economic disincentives for assessing negative externalities, as much as the problems of attaining consensus on social costs and benefits, mitigates against sole reliance on corporate accountability and assessments of developing technologies.

Although the presumption that the private sector should do technology assessment is contentious, it nonetheless underlies debates over the role of and responsibilities for technology assessment. Surveys of the use of technology assessment in industry also tend to presume that the private sector does or should conduct such assessments.[9] The tenuous nature of this assumption means that much of the information about the conduct of technology assessment in the private sector is questionable and possibly misleading. On the one hand, a good number of private-sector decision makers do not understand what is entailed by technology assessment. We have found, for example, that some representatives of private industry tend to label as technology assessment everything from legally required safety checks to purely profit-motivated market research. On the other hand, some chief executives simply deny the relevance and propriety of technology assessment in the private sector.

Many corporations do conduct analyses that are analogous to public-sector technology assessment. Yet such analyses are inherently different in focus and scope because they are intended to support corporate decision making. While they may be future oriented and focused on the potential consequences of technological initiatives, their objective is to enhance the long-range viability of industry rather than to assess societal benefits and costs. Private-sector studies typically include, for example, analyses intended to help corporate managers ensure that a firm complies with the regulatory restraints, anticipates restraints or liabilities that might be imposed in accordance with evolving public policies, and identifies changing societal demands in order to evaluate future opportunities and constraints on corporation activities.

Private-sector studies usually focus on the impact of social change on industry rather than on the impact of the industry on society. In some companies a form of social-benefits accounting—usually known as the social audit—has been developed. Still, social audits are basically only an extension of the annual financial report to stockholders. They merely attempt to display the societal benefits such as employment, participation of employees in community-service activities, additional tax revenues, and contributions to charitable and educational programs created by a company and by its location in a particular community or region. The Securities and Exchange Commission (SEC) recently promulgated a requirement that "public companies" disclose environmental costs associated with their activities in order to protect stockholders from environmental liability.[10] Some companies have therefore begun to conduct so-called environmental-compliance audits of their activities and facilities in order to assure themselves and their stockholders that they are meeting federal standards aimed at protecting the environment. Engineering and accounting contracting firms, recognizing this as a marketable new service, are also encouraging their clients to have an environmental-compliance audit. At least three corporations—U.S. Steel, Allied Chemical, and Occidental Chemicals—have initiated or carried out these environmental audits.

Technology assessment in the private sector, on the whole, appears to be rare and infrequent. At best, firms in the private sector undertake a range of analyses that are analogous to, but not identical with, technology assessments undertaken by the government. In order to better understand the extent to which technology assessment and related activities are undertaken by industry, we conducted a literature search for topics related to industrial technology assessment. We also interviewed twenty-five corporate executives who had attended technology-assessment workshops, since we believed these people would be the most familiar with the concept within their firms.[11]

Although our sample was small and by no means representative of all industries, two preliminary conclusions are suggested by the survey. First, technology assessment does not appear to be a commonly understood concept in the private sector even in these corporations. Second, the term *technology assessment* was employed to denote a variety of wide-ranging activities. There were obvious instances of "old wine in new bottles" as marketing studies and feasibility studies were relabeled technology assessments. The survey shows that technology assessment or, more precisely, parallel activities in the private sector should be clearly differentiated from public-sector technology assessments.

The Private Sector and Technology Assessment

The interviewees were already familiar with technology assessment by virtue of having attended at least one short course or conference on the topic. However, we were regularly told by interviewees that they were the only persons in their firms who had substantial interest in technology assessment. One corporate official, who had spent ten years as a consultant to *Fortune* 500 companies, claimed that he had never encountered technology assessment in any of the companies he worked with. Technology assessment, he said, was in the position in which strategic planning found itself twenty years ago; he guessed that maybe one in a hundred corporate officers have even heard the term.

Several factors contribute to the lack of concern and knowledge about technology assessment in the private sector. First, the idea goes back barely over a decade and originated in the field of public policy. Second, we found that technology assessment has been infrequently considered in industry trade presses. Magazines such as *Fortune* and *Forbes,* as well as the more-specialized business newsletters, largely ignore technology assessment. Discussions of technology assessment appear in articles primarily directed to the scientific and engineering communities or to the public-policy community. Those chief executives who were the most aware of technology assessment, we found, were in companies primarily involved with advanced technology or enterprises with a high degree of public and governmental visibility.

Respondents revealed an interesting split in their understanding of the complexity of technology assessment. Some individuals tended to view it as a commonsense technique that was widely used but under different names by a variety of companies. As one person commented: "I believe any major company, of necessity, plans with tools like technology assessment without ever giving them special names." Another felt that it was a fairly straightforward, commonsensical approach to dealing with emerging technologies and that people in a number of contexts were doing technology assessment without calling it such. Conversely, some respondents considered technology assessment to be too elaborate and rigid a technique to be used by most corporations. At the extreme was one rather cynical executive from a large company who viewed technology assessment as an elaborate process designed to make a great deal of money for consultants. He explained his company's lack of any technology-assessment activities by saying, "We voted with our feet; we feel that as a concept, technology assessment is not relevant.''

Individuals indicating the most familiarity and interest in technology assessment were employed by companies that had several characteristics in common. For example, the companies were usually engaged in high technology with a relatively large proportion of their work force trained in some field of engineering or the sciences. In these companies, executives tended to see their activities as having a direct effect on the environment or a major impact on society. Perhaps more important, they also tended to perceive themselves as especially vulnerable to pressures from outside of the company, particularly from special-interest groups or from governmental regulations. Many of these executives viewed technology assessment solely in terms of anticipating the effects of the political and economic environment on their activities, rather than examining the socioeconomic consequences of their firm's products or activities. One chemical-company executive stated this succinctly in explaining the technology-assessment activities of his corporate-research division: "Research managers spend a great deal of time looking at future trends and trying to identify future opportunities for a high-technology chemical company."

Those executives having the least interest in technology assessment worked for companies involved in the manufacture and distribution of basic consumer goods such as food products and household items. Respondents from these corporations had largely given up any idea of instituting some form of corporate technology assessment, even if they personally considered it a good idea. A marketing executive from one large food company admitted that the prospect for corporate assessment was not "terribly realistic or likely" in his corporation. Yet, another executive from a multi-product corporation felt that something like technology assessment might be useful if they ever decided to introduce a product that was radically different or controversial.

A fair number of firms fell in between these two categories. In these firms we found individuals familiar with, but also ambivalent about, the prospects for technology assessment. While the management of the companies was not opposed to technology assessment, it was not particularly interested in the concept either. A research engineer with a high-technology manufacturing firm—one that produced no consumer products—expressed the idea simply: "They feel they'll get around to looking into it sooner or later." Another person involved in technology forecasting indicated that one of the main factors that could spur private-sector interest in technology assessment was a governmental requirement that it be done.

Essentially, the survey indicates that technology-assessment concerns and activities are quite limited in the private sector. Even in those corporations in which interviewees claimed to be doing technology assessment, the technology-assessment-related activities were modest. In no instance did we find a corporation with an established technology-assessment process that appeared to be in any way crucial to its decision-making process.

Industrial Technology Assessment

Although technology assessment appears neither understood nor valued in the private sector, some firms nevertheless label a broad range of activities technology assessment. Firms have different, and often conflicting, definitions of the term. Within a single firm we found individuals to be working on the basis of conflicting definitions. For instance, two respondents—one involved in technology forecasting and the other a corporate planner— within the same branch of the same firm appeared to have mutually incompatible conceptions of technology assessment. The technology forecaster maintained the firm was not doing any technology assessment and saw no realistic possibility of its doing any in the near future. He felt that management did not consider such studies cost-effective. His colleague, a corporate planner, however, stated that the corporation was doing technology assessment because management considered it important. The studies he referred to as technology assessments appeared to be forecasts of future technologies and their potential for the corporation, and they were undoubtedly done by the forecaster who did not consider these studies technology assessments.

Respondents frequently listed the following activities as technology assessment: technology forecasts, market analyses, engineering evaluations, site analyses, environmental scans, competitive analyses, and economic and business projections. While some of these activities do contain elements of technology assessment, it was also apparent that respondents frequently simply explained a company's regular planning process and marketing strategies under the rubric of technology assessment.

A large number of respondents identified market analysis as technology assessment, or the closest thing to it, done by their firms. In most cases, the person was aware that such studies do not fit a definition of technology assessment very well. As one company executive said, "The closest thing to technology assessment [in our firm] is . . . trying to predict areas of business opportunity for the future." At the same time, he admitted that the company did "very little in the way of analyzing the effects of their business activities on either society or the environment."

Several large, technology-dependent corporations had offices or divisions that regularly engaged in activities that could be viewed as inverted technology assessments. These divisions were concerned with anticipating future societal developments for the purpose of analyzing their potential effects on the firm. One corporation established a twofold approach to this form of industrial technology assessment. It had a small, centralized futurist group reporting to the executive management and a shorter term project and planning office within each of its several technology areas. The smaller futurist group had been begun by a group of people within the firm. The group informally looked at a number of techniques potentially useful for long-range planning. Subsequently, the futurist group was established as a permanent division within the company, reporting directly to the top levels of management. The division conducts its own studies, produces reports for management, and conducts brainstorming sessions within the company on topics such as coping with diminishing sources of traditional energy. While they focus on future technological developments and economic and regulatory trends, they do not deal with the social implications of corporate activities. The company's second route for industrial technology assessment involves planning offices within its respective technology areas. These offices focus on areas of production and broad corporate concern such as energy and natural resources; but again, their primary orientation was toward socioeconomic impacts on the firm.

A number of respondents either identified specific instances of technology-assessment-related studies—for example, environmental-impact studies—or indicated that their companies would, under certain circumstances, be open to conducting one-time, ad hoc technology assessment. However, a researcher in one electrical company related an interesting account of an unsuccessful technology assessment that left management leery of undertaking any further ones. The company had contracted a consultant to assess the effects of rate structures on electricity consumption and their social and regulatory consequences. The company felt that the study fell far short of expectations and that they had been promised much more than was delivered. When the researcher at a later point proposed a much more-limited, in-house technology assessment in his own area of expertise, he met considerable resistance. The general reaction was, "This looks interesting, but we don't want to put the time and effort into it." Management apparently considered technology assessment to be of dubious utility. By contrast, we did find in a major national R&D corporation a defense-related group conducting a technology-assessment-related study—namely, an environmental-impact statement for the MX-missile program. This was a relatively large project performed by both in-house staff and outside consultants and was notably funded by the U.S. Air Force.

In sum, we found several fairly typical responses to our questions about private-sector technology assessment. First, technology assessment, even in its inverted form (namely, assessment of the impact of socioeconomic forces on the firm's activities), seems to be hard to sell to management.

Second, assessments of the social and ecological impact of a firm's activities will usually be done only to the extent necessary to satisfy regulatory requirements. As one respondent wrote:

> Because of the nature of our business, we normally have no requirements to perform assessments of our technology developments. That is, being prominently in the space, communications, and software businesses for the government, we do not become involved with justifying what we do. . . . However, from the technology futures point of view, studies of where our business might go and what problems might be encountered, I feel that TA studies would definitely be helpful. But, until management is incentivized [sic] for the long term, support for such work is unlikely unless funded by the government.

When respondents gave examples of corporate assessments of broader social issues, those analyses usually paralleled a demonstrated governmental interest in those issues. One chemical-company executive stated that wherever there is a government policy, the company tries to formulate its own response and/or analyze its activities in terms of that policy—for example, environmental policy.

Finally, aside from these responses, several individuals mentioned activities that might be called informal technology assessment. In some instances this was actually much closer to the concept of technology assessment as a tool for identifying the second-order impacts of a technology than many of the more-formal studies cited. Informal technology-assessment activities refers to individual or small-group efforts to monitor technological changes and to study their possible societal impact. The information was used in reports to top management without being called technology assessment, in part because the individuals had concluded that their corporations would not support a formal program of technology assessment. Comments like the following reflect the nature of this informal and unsystematic process of technology assessment:

> We do try to bring in social effects, but this isn't a primary thing. It is done on an informal basis. As issues are raised, we try to get the information to the highest level. I've tried to familiarize myself with the issues and to inject this knowledge of social and political factors into the company's decision making. Another comment: We try to keep our eyes open and sense emerging trends, but there is no formal group doing technology assessment.

Technology Assessment as Public-Policy Analysis

Industrial technology assessment involves a different set of activities from those identified with public-sector technology assessment. This is not undesirable, but the two activities and their objectives should not be confused.

Industrial technology assessment, even in its most sophisticated form, is fundamentally an analysis of the effects of technological, socioeconomic, and political change on the industry doing the assessment. Only secondarily, if at all, are industries concerned with how their activities affect the external environment, let alone with assessing the social costs and benefits of their activities. The largest volume of private-sector technology-assessment-related activities undoubtedly involves testing of new drugs by pharmaceutical companies and assessments by industries of their environmental impact. Yet, these are also the areas in which corporations have been increasingly required by the government to analyze the impact of their activities. Few incentives exist within the private sector to conduct further analyses of the social and environmental effects of technological and industrial developments. More-rigorous analyses would involve committing resources to collection of information that most chief executives feel they do not need. As one respondent stated, "Corporate officers tend to get information that helps them make specific decisions or deal with specific problems. These decisions are, first, financial and, second, technological. Secondary and tertiary impacts are generally discounted by the corporate world." Corporate executives typically do not even value inverted technology assessment, although planners understand that negative externalities can come back to haunt a corporation in the form of public pressure for more government regulation. When chief executives do appreciate the import of studying the externalities of an industry, assessments are occasionally directed toward identifying ways of manipulating public opinion and avoiding or manipulating government regulation.

What is necessary, then, is an explicit acknowledgment of the intrinsic differences between technology-assessment objectives cum activities in the public and private sectors. Private-sector assessments should not be considered substitutes for technology assessment in the public sector. One simply cannot assume that private-sector assessments of technological options will either be objective or will acceptably balance the social costs and benefits of developing technologies. Technology assessment belongs in the public sector. To the extent that public-spirited corporations are willing to take on the task of doing technology assessments, they could contribute to more-effective and -efficient implementation of public policy.

Firms in the private sector might also benefit by avoiding costly mistakes and by identifying profitable and socially advantageous courses of action. In any event, the point remains that to use the term *technology assessment*—a term that, whatever its own shortcomings, has been firmly attached to public-policy analysis—to describe parallel but fundamentally different corporate-planning activities is a mistake. It both obscures the current status of these activities and creates barriers to their future development and improvement.

Notes

1. See, for example, "Technology Assessment Seeks Role in Business," *Chemical & Engineering News* (28 March 1977):11-13; Harvey Brooks and Raymond Bowers, "The Assessment of Technology,"*Scientific American* 222 (1970):15-16; and Harold Green, "Limitations of Implementation of Technology Assessment," *Atomic Energy Law Journal* 14 (1972):62.

2. See, for example, William King and David Cleland, *Strategic Planning and Policy* (New York: Van Nostrand Reinhold, 1978); Michael Moskow, *Strategic Planning in Business and Government* (New York: Committee on Economic Development, 1978); and Albert C. Worrell, *Unpriced Values: Decisions without Market Prices* (New York: John Wiley, 1979).

3. Carl Friedrich, "Political Decision-Making, Public Policy and Planning," *Canadian Public Administration Review* (1971):144.

4. See statements of Emilio Daddario on technology assessment, in U.S. Congress, House, Committee of Science and Astronautics, Subcommittee on Science, Research, and Development, Committee Print, 90th Cong., 1st sess., 1967. See also U.S. Congress, House, Committee on Science, and Astronautics, Subcommittee on Science, Research, and Development, *Proceedings,* 90th Cong., 1st sess., 1967; and U.S. Congress, House, Committee on Science and Astronautics, Subcommittee on Science, Research, and Development, *Hearings on Technology Assessment,* 91st Cong., 2nd sess., 1969.

5. For further discussions, see Alan L. Porter, Frederick Rossini, Stanley Carpenter, and A.T. Roper, *A Guidebook for Technology Assessment and Impact Analysis* (New York: Elsevier North Holland, 1980); S.R. Arnstein and A. Christakis, *Perspectives on Technology Assessment* (Jerusalem: Science and Technology Publishers, 1975); Joseph Coates, "Technology Assessment—A Took Kit," *Chemtech* (June 1976):372; J. Coates, "The Role of Formal Models in Technology Assessment," *Technological Forecasting and Social Change* (1976):139; Vary T. Coates, *Technology and Public Policy: The Process of Technology Assessment in the Federal Government,* vols. 1 and 2 (Washington, D.C.: George Washington University, Program of Policy Studies in Science and Technology, 1972); and V.T. Coates, *Technology Assessment in Federal Agencies, 1971-1976* Report to National Science Foundation, NTIS #PB-2925969/OSL, A25, 1977).

6. Milton Katz, "Decision-Making in the Production of Power," *Scientific American* 223 (1971):192.

7. Emilio Q. Daddario, "Technology Assessment Legislation," *Harvard Journal on Legislation* 7 (1970):516.

8. For a further discussion, see David M. O'Brien, "The Courts, Technology Assessment and Science-Policy Disputes," chapter 6 of this

book; Arthur Pigou, *The Economics of Welfare,* 4th ed. (London: Mcmillan & Co. 1932); R.H. Coase, "The Problems of Social Cost," *Journal of Law and Economics* 3 (1960):1; James Buchanan and William Stubblebine, "Externality," *Economica* 29 (1963):371; and Ralph Turvey, "On Divergences between Social Cost and Private Cost," *Economica* 30 (1963):309.

9. See, for example, "A Survey of Technology Assessment Today," [Report prepared by Peat, Marwick and Mitchell & Co., (M. Breslow, principal investigator) for the National Science Foundation, 1972]; (NTIS #PB–221850, A05, 1972); J. Maloney, "Technology Assessment in the Private Sector" (Report prepared for the National Science Foundation by Midwest Research Institute, 1979); J. Maloney, "Technology Assessment in the Private Sector: Some Findings of Potential Use to OTA" (Draft report for the Office of Technology Assessment, (Midwest Research Institute, 1981); U.S. Congress, Office of Technology Assessment, *Technology Assessment in Business and Government: Summary and Analysis* (Washington, D.C.: U.S. Government Printing Office, 1977); and U.S. Congress, Office of Technology Assessment, *Hearings, Technology Assessment Activities in the Industrial, Academic, and Governmental Communities* (Washington, D.C.: U.S. Government Printing Office, 1976).

10. Securities and Exchange Commission, 17 C.F.R. 229.001, Item 5, Instruction 5. In May 1981, the SEC proposed an amendment to limit the required disclosure to matters that would potentially subject the company to penalties or sanctions of $100,000 or more. The amendment would eliminate the need to disclose, for example, routine violations of clean-water standards where the penalty was only a few thousand dollars. The final rule is expected about January 1982, after public comments have been evaluated.

11. The interviewees requested and were granted anonymity. Three persons could not be located, and two additional responses came from persons recommended by one of our initial interviewees.

Technology Assessment in the National Government

Vary T. Coates

The National Political Process and Technology Assessment

During the 1970s, a subtle but profound change occurred in federal-agency policymaking processes. The factors regarded as pertinent in formulating public policy, in designing projects and programs, and in allocating funds became broader and more complex. Administrators began to examine social as well as economic costs and benefits and to explore the secondary or indirect impacts as well as the primary objectives of public policies. In short, the concepts of strategic planning began to permeate administrative decision making and policy formulation.

This change in administrative procedures and practices was largely forced on federal agencies by a combination of congressional and judicial pressures. It progressed slowly, incrementally, and unevenly; some agencies remained unaffected. Yet, those agencies dealing with science and technology and with large public works were, almost without exception, forced to consider the possible socioeconomic consequences of their programs and projects and to justify those programs and projects in new terms that went beyond traditional categories of design objectives and direct budgetary costs.

Technology assessment as a formal analytical approach or coherent institutionalized activity was only one factor in the changes in agency decision making and policy formulation. For a brief period, *technology assessment,* in a strong and narrow sense, became one of those fashionable buzzwords in Washington, D.C., around which national and international conferences are organized. Technology assessment in a weaker and broader sense is a class of policy-oriented research or policy analysis.[1] As such, it aims at identifying and evaluating the immediate and secondary impacts of a technology, industrial process, or large-scale public-works project.[2] While analyzing issues of technological feasibility, direct or internalized costs, and primary benefits, technology assessment is distinctively and primarily concerned with the identification and evaluation of indirect secondary impacts. Technology assessment therefore subsumes social, economic, environmental, legal, institutional, and cultural impact assessments. In most cases,

33

technology assessment is interdisciplinary social-science research, including qualitative and quantitative data.

Technology assessment in the strong and narrow sense had its day and has faded into the background along with planning by objectives, and Program-Planning and Budgeting Systems (PPBS), and other nonce phrases cum slogans recorded in the history of public administration. Technology assessment in the weaker, broader, and more-important sense is likely to have greater and more-lasting political significance. The basic questions that have been raised about technology and democratic governance in the last decade will undoubtedly persist. Moreover, uncritical public acceptance of simple technological answers to complex social problems can no longer be assumed. After a series of technological scares and a few disasters such as the controversy over rDNA, nuclear power, Love Canal, and so on, and after a decade of public participation and activism on the local and national levels, the general public remains concerned about and committed to environmental protection and natural-resource allocation as well as wary of both the promised benefits of technology and governmental programs to develop and use advanced technologies.[3]

Before examining the evolution and current status of technology assessment in the national government, it bears emphasizing that technology assessment developed in three quite different political contexts and institutional settings: (1) the NSF, (2) the federal agencies of the executive branch, and (3) the OTA. The differences in the shape and scope of technology assessments in the national government derive primarily from the differences in the scope of the responsibilities, powers, and authority of these respective institutions. In all three institutional contexts, however, the movement to formalize technology-assessment activities slowed after 1975. Nonetheless, the primary aims of technology assessment have become generalized and widely accepted (albeit not as widely practiced) within the national government. Indeed, technology assessment in the federal agencies remains, for the most part, sporadic, narrow in focus, and disjointed. By contrast, the technology-assessment program of the NSF and the OTA have remained more consistent; although, as is discussed later, they too lost some of their momentum during the late 1970s and early 1980s.

The impetus for technology assessment never came from the highest levels of the executive branch. Presidents and their advisors usually prefer to preserve their options and not to be limited by too much public information about the social costs or potential risks that inevitably accompany political action. The subtle pressure that stimulated the gradual and often reluctant acceptance of technology assessment came from social and political forces outside of government and from inherent tensions that operate among the branches of government and among federal agencies and their constituencies.

In 1966, Congressman Emilio Daddario, then chairman of the Subcommittee on Science, Research, and Development of the House Committee on Science and Astronautics, introduced the concept of technology assessment, proposing the establishment of an office of technology assessment to serve Congress.[4] Congressman Daddario's proposal occurred at a time of growing public alarm over alleged health, safety, and environmental hazards resulting from chemical and industrial processes and from the unexpected side effects of consumer products.[5] Moreover, urban-redevelopment and highway and airport projects were causing people to be relocated and communities to disintegrate, which thus brought protests, demonstrations, and citizens' lawsuits. Congress was shaken by the idea that once popular pork-barrel projects seemingly bode profound political protests. Indeed, Congress became intensely suspicious of the planning and policymaking processes of executive agencies. Within this political climate Congressman Daddario proposed an office of technology assessment to give Congress early warning of the undesirable consequences that might flow from new technologies and federal projects. Technology assessment thus was perceived as both a step forward to more-democratic policy formulation and a weapon that Congress might employ when challenging executive-branch programs and projects.

In the late 1960s, the House Committee on Science and Astronautics took the unusual step of commissioning studies and evaluations of the idea of technology assessment by the National Academies of Science, Engineering, and Public Administration. The idea of technology assessment also quickly gained acceptance among professionals and academic specialists in science-policy research. In 1969, the NSF initiated a program that was the real progenitor and developer of technology assessment. The OTA was finally established for Congress in 1972, with its operations beginning in 1973.

The NSF Technology-Assessment Program

The NSF was created in 1950 to promote the progress of science through the support of basic scientific research and education. Research grants were typically made along disciplinary lines (in response to research proposals from an individual investigator, usually at a university). The two main objectives of NSF remain: (1) advancing the state of basic scientific knowledge and (2) improving the quantity cum quality of scientific education.

In the 1960s, however, the NSF expanded its objectives in response to congressional pressures for studies of social and environmental problems such as poverty, urban renewal, civil rights, and environmental issues. In 1967, the NSF organized a new program entitled Interdisciplinary Research Related to Problems of Our Society (IRRPOS). This program was later placed with a new directorate of research applications and given the title of Research Applied to National Needs (RANN). Not coincidentally, NSF's oversight committee in the House of Representatives was the Science and Astronautics Committee, which was also studying Congressman Daddario's proposal for an office of technology assessment.

The NSF program deviated from other established NSF programs in several ways: It contracted out for interdisciplinary projects that were conducted by private organizations including private industry, and it did not rely solely on unsolicited proposals but rather initiated research by contracting with experts and by funding proposals through competitions based on merit. The NSF program had four broad objectives: (1) performance of technology assessments through outside contracts, (2) support of scientific research required for or implied by such assessments, (3) sponsorship of related conferences and symposiums, and (4) preparation of in-house position papers and policy recommendations.

The two guiding principles of the NSF technology-assessment program—that is, contracting out for interdisciplinary studies and an active NSF role in selecting problem areas and soliciting proposals—were in some ways contrary to other established NSF policies and procedures. This, in turn, had a powerful affect on the NSF technology-assessment program. Technology-assessment activities never became prominent within NSF's structure and budget. In fact, the technology-assessment program was only one component of the IRRPOS program and by 1969 was placed in the smallest division of RANN, the Office of Exploratory Research and Problem Assessment. In 1972, the NSF director, Guyford Stevens, was also made science advisor to President Richard Nixon. Yet, when he established a separate policy office within the NSF to serve him, representatives of the technology-assessment program, although the most policy-oriented program in the NSF, were not included in that office. Thus, the technology-assessment program, from its conception, was never prominent within NSF's budget or organizational structure. The program received 3 percent or less of the RANN budget until the end of the RANN program in 1978, ranging from a low of $667,803 in FY1973 to a high of $1,433,174 in FY1975.

The four main objectives identified with the NSF program had only mixed success. Other NSF directorates largely ignored the technology-assessment program, and their programs continued to fund mostly researcher-initiated projects. Profit-making research organizations received a large number of contracts along with nonprofit research centers and interdisciplinary university-based research groups. The research organizations and some of the university groups later offered similar research services to federal agencies and, in general, spread the concept of technology assessment through government. During the early years, NSF staff also played a fairly active role in shaping the research agenda, identifying technologies where assessments were needed, soliciting research proposals, and working with recipients to shape useful reports. Still, there was relatively little NSF support for technology-assessment conferences and symposiums, although such meetings were, in the mid-1970s, an important factor in the growing acceptance of technology assessment. Preparation of NSF in-house policy papers and recommendations based on technology assessments were neither encouraged nor allowed except when necessary for presentation to congressional hearings.

In spite of these problems, the NSF program did contribute to a developing consensus about the definition, scope, components, and techniques of technology assessment. From 1971 through 1980, the NSF funded about fifty-two technology assessments and several-dozen additional studies, which included analyses of technological assessments conducted by governmental agencies, a few symposiums and conferences, and stocktaking studies to evaluate NSF-funded assessments. Since the congressional OTA did not begin operations until 1973, and since technology assessments undertaken by federal agencies were few, the NSF program was the only source of technology-assessment funding during the early and mid-1970s.

The NSF notably neither has responsibility for development and promotion of specific technologies or industries as, for example, does the National Aeronautics and Space Administration, nor responsibility for controlling or regulating technologies as does the EPA. Hence, a major source of institutional bias that afflicts mission-oriented agencies when they undertake technology assessment was not present at the NSF. The corresponding problem, however, arose that the NSF was constantly in the position of committing the political sin of intruding on the turf of other agencies, which in turn left it vulnerable. Accordingly, the NSF adopted the strategy of concentrating on technologies that either did not fall within any agency's responsibility or for which several agencies had partial responsibility. In a few cases, NSF program managers were also able to persuade agencies to undertake their own technology assessments or to share with the NSF the funding and management responsibility for assessments. The NSF thereby avoided political conflicts and helped substantially in achieving acceptance of technology assessments by federal agencies. The NSF technology-assessment program also covered a remarkably broad range of subjects. Of the substantive studies funded, some were problem-oriented, such as food and waste disposal, automobile-derived pollution, energy conservation, and freon in the atmosphere. The more-technology-oriented studies covered a range of industrial technologies, biological and medical technologies, as well as social/institutional technologies.

In retrospect, the NSF program moved through several definite phases. Initially, between 1971 and 1974, the emphasis at the NSF was on experimentation and on producing a body of technology assessments in a wide range of subject areas. Researchers were encouraged to contribute to the development of a coherent and consistent body of knowledge and, at the same time, to experiment with diverse techniques and methods. Public dissemination and discussion of research results was encouraged; in fact it was demanded.

During 1975–1977, a movement away from assessment of single emerging technologies occurred. Greater emphasis was placed on funding research in clusters of closely related technologies and on methodological studies. The program concentrated on functionally defined technologies such as life-extending technologies, human-rehabilitation measures, quality-of-work-life schedules, and transportation-telecommunication trade-offs. Several studies dealt with technological advances in information sciences, sanitation and waste-water treatment, and modern corporate-management techniques. The definite methodological purpose of these studies was to aim at developing an understanding of the relationship between technological and social change by historically tracing the impacts of particular techniques.

Funding for the technology-assessment program from 1978–1981 was divided to cover a closely related program in risk analysis. There was also an emphasis on reevaluation of assessment methods and techniques, with a concentration on computers and telecommunications technologies. In addition to the changing focuses of the NSF program were significant organizational changes. In 1978 the RANN program was dissolved. The technology-assessment program was moved to the Policy Research and Analysis Division of the directorate for Scientific, Technological, and International Affairs (STIA). The president's science-policy advisor and his council were removed from the NSF and located within the White House. STIA was reorganized and became dominated by the personnel that had previously made up the NSF's national research-and-development-assessment program, which had focused on industrial innovation and technology transfer, concentrating on short-term, economic benefits of technological development and the use of quantitative techniques—especially cost-benefit analysis. The technology-assessment program, with its emphasis on social science and analysis of indirect and secondary impacts, thus remained of secondary importance within its own directorate.

These organizational changes significantly affected the operation of the NSF program. The program indeed appears to have undergone further retrenchment from 1979–1981. The bulk of technology-assessment funding has been allocated to methodological and reevaluation, or stocktaking, studies, which also are usually noncontroversial. Furthermore, substantive assessments have been predominantly directed at computer and telecommunications technologies, which offer a rich field for social analysis but are not regarded as within the preserve of any powerful federal agency; thus they also tend to be noncontroversial.

The NSF technology-assessment program suffered throughout its ten-year history from several more-serious problems. Research teams who performed well in interdisciplinary research were not always skilled at policy analysis. The idea of interdisciplinary analysis is itself elusive and ambiguous. Moreover, the primary constituency of the NSF has always been scientists engaged in basic physical and biological science and, secondarily, research engineers. These groups tended to be suspicious of interdisciplinary (or nondisciplinary) studies. Basic-science researchers were concerned that emphasis on useful and policy-oriented research would both reduce funding for traditional scientific research and violate the professional canon that scientists, not the government, determine the direction of research. Technology assessment, therefore, frequently did not have strong supporters within the NSF.

The vulnerability that resulted from the lack of a large, well-established constituency and of strong support by the NSF upper management, as well as the organizational changes, perhaps rendered inevitable the retrenchment apparent in the NSF program. The NSF remains nonetheless the potentially most promising environment within the federal government for conducting comprehensive and exploratory technology assessment. Given a commitment to sponsor even-handed, future-oriented policy evaluations of emerging technologies and to make them available to informed and concerned publics, the NSF could emerge as an auspicious arena for technology assessments since it remains isolated from most of the political pressures encountered by federal agencies and also from the pressures of congressional committees that are placed on the OTA.

The Federal Bureaucracy and Technology Assessment

The social forces that brought about the introduction of technology assessment in the legislative branch and in the NSF also operated on the executive branch. Public disillusion with the benefits of technology and the increasing skepticism about governmental programs had a direct effect on agency-supported projects. For instance, the National Environmental Policy Act (NEPA), passed in 1969, provided interest groups with a formidable weapon against unwelcomed government projects.[6] During the first two years after enactment of the NEPA, the courts in no less than nineteen instances issued injunctions delaying federal action (or a state action with federal funding) in order to achieve compliance with the act.[7] Since agencies increasingly had to defend their regulations in judicial arenas, they became more receptive to assessing the impact of developing technologies and of their programs and projects. Science- and technology-oriented agencies became particularly interested in the concept of technology assessment.

However, many barriers exist to the development of technology assessment within the federal bureaucracy. The White House and the Office of Management and Budget (OMB), as previously noted, never pushed development of technology assessment and, in the late 1970s, repeatedly cut funds for exploratory impact studies.[8] Legislative mandates, potential conflicts between assessments and the primary objectives of the agencies, the demands of institutional survival, and the lack of time, funds, and personnel for comprehensive studies further discouraged federal agencies from undertaking technology assessments. Some agencies have a responsibility for promoting one or more technology or industry—for example, the Federal Aviation Administration (FAA) has regulatory responsibility for developing a domestic air-transportation system. Yet typically, federal responsibility for monitoring and developing technologies is split among many agencies. Pesticides, for instance, are governed by the Department of Agriculture (DOA), EPA, Food and Drug Administration (FDA), the Department of Interior (DOI), and to a lesser extent, several other agencies. Federal agencies also usually have multiple and even conflicting responsibilities. The DOI, for example, repeatedly faces internal struggles related to its multiple responsibilities for natural-resources development, conservation and management of public lands, and trusteeship over Indian lands. The DOA, likewise, must be responsible to agribusiness, small family farms, consumers, and commodity markets. Thus, comprehensive impact assessments threaten to generate internal conflicts within an agency and between the agency and its various clientele. While technology assessments may inform decision making, they may also invite political conflict since they often pose benefits for one or more constituency while indicating added costs for others.

In spite of these obstacles, federal agencies did begin in the late 1960s to give consideration to the secondary impacts of developing technologies and public-works programs. During the 1970s, a marked change occurred in the way in which decisions were made about research funds, support for technological development, civil-works projects, and even to a lesser degree, technology regulation. Several large studies were done by agencies well before either the NSF technology-assessment program or the OTA were in operation (although none was labeled technology assessments).[9] Federal-agency decision making during the late 1960s and early 1970s increasingly gave greater attention to the possible indirect, unplanned, secondary impacts of governmental programs related to science and technology.

Federal agencies actually conducted what might be appropriately called secondary-impact analysis. Secondary-impact analysis is not the same as comprehensive technology assessment. Administrative responsibility for impact analysis is usually dispersed through an agency. Consequently, analysis of secondary environmental impacts is done in one office, analysis of secondary economic impacts is done somewhere else, and analysis of social, legal, and institutional impacts is conducted in still other offices. Technology assessment, by contrast, aims to integrate all available information and insights into one study and thereby to lay out for decision makers the factors that should be considered, the policy issues posed by a particular technology, and the range of policy options.

When the first survey of technology assessment in federal agencies was completed in 1972, many federal administrators and planners said that they "took into account some of the secondary consequences of technological applications," and some reported that analysis of such secondary impacts was their major activity.[10] No agency, however, had yet institutionalized a continuing technology-assessment program. Only six studies undertaken by federal agencies from 1966–1971 were identified as comprehensive technology assessments. (None of these was identified as technology assessments when they began, but two were so described in the agencies' final reports.)

By contrast, between 1971 and 1977, at least five major agencies formally established technology-assessment programs. In 1972, the Department of Transportation (DOT) was the first cabinet-level department to institutionalize technology-assessment capabilities when the decision was made that all transportation systems would be subjected to a comprehensive socio-economic impact assessment. Within the DOT, the Office of the Assistant Secretary for Policy Planning and International Affairs undertook several three-to-five-year studies, such as the Bay Area Rapid Transit (BART) Study Program and the Climatic Impact Analysis Program (stratospheric aircraft). The DOT also cosponsored, with the NSF and the National Aeronautics and Space Administration (NASA), assessments of intercity and large-scale aircraft-transportation systems. Because the term *technology assessment* was new and not widely familiar in 1972, the DOT program was not formally called technology assessment but rather systems analysis.

The DOA, after an exploratory technology assessment of minimum tillage in 1974, convened a staff workshop on technology assessment in 1975. This led the DOA subsequently to establish a technology-assessment program called TIFFS—Technology Innovation in the Food and Fiber Sector. TIFFS conducted assessments of energy from biomass, automated food retailing, solar-energy applications for agriculture, and large-tractor technology. These assessments were designed to support agency policy formulation, to improve the allocation of R&D funds, and to inform constituents about impending technological changes.

The FAA established a technology forecasting and assessment group within its Office of Aviation Policy in 1975. The group first produced a forecast and exploratory assessment of minicomputers on aircraft. The office was to begin with technological forecasting and move gradually toward comprehensive impact studies, but the latter was quietly dropped from the agenda in 1978.

The EPA also initiated several technology assessments between 1972 and 1974, which were eventually transferred to the new Energy Research and Development Administration [ERDA, later the Department of Energy (DOE)]. In 1974, an OMB-sponsored interagency task force on health and environmental effects of energy use issued the King-Muir report, which called for an impact assessment of all proposed energy-development projects. The EPA then established an integrated technology-assessment program (ITA). The ITA program initiated five multiyear, multimillion-dollar contractor assessments of the Western Energy Development, the Ohio River Valley Energy Development, the Appalachian Energy Development, electric utilities, and coal conversion.

Unlike most of the other technology-assessment programs in federal agencies, the EPA program (now in the Office of Strategic Assessments and Speical Studies of the EPA) has continued to fund technology assessments, primarily small exploratory assessments called mini-assessments. Ten mini-assessments were completed from 1980–1981, and another seven or eight were funded for 1982. The program funded studies, for example, on bromines, telematics (computers and telecommunications), increased use of wood burning, hazardous facilities, applied genetics, composite materials, alcohol fuels, and alternative chemical feedstocks. In addition to its mini-assessments, the EPA undertook a large-scale technology assessment of energy development in the Sunbelt regions.

The ERDA inherited from the EPA several ongoing technology assessments and some staff. Accordingly, the ERDA established a technology-assessment branch within its division of transportation energy conservation, which continued the EPA assessments and initiated technology assessments of the electification of transportation systems. In 1977, the ERDA also established an ITA program within the Office of Environment and Safety. The title of both the program and the office has changed several times during the ERDA's and DOE's troubled history. Presently, the program is located in the DOE's Office of Environmental Assessment.

The Office of Environmental Assessment concentrates chiefly on studies of energy technologies and on development of a complex mathematical model, the strategic environmental-assessment system (SEAS). The office nevertheless has conducted some broader studies, notably a technology assessment of solar energy and assessments of the socioeconomic impact of energy development in several regions of the country. In 1982, with President Ronald Reagan's administration considering dissolution of the DOE, the continuation of this program (and indeed the Office of the Assistant Secretary for Environment) remains in doubt, even should the DOE survive.

Finally, during the late 1970s, technology-assessment programs and activities were established in other agencies to examine the impact of industrial and technological developments on public health and safety. The Department of Health, Education, and Welfare, [DHEW, now the Department of Health and Human Services (DHHS)] began, in the mid-1970s, planning development of technology-assessment capabilities within its National Center for Health Services Research. The DHEW formally established a departmental mechanism for assessing medical technologies, but at some indeterminate period in the transition from the Carter to the Reagan administration, this innovation quietly faded away. The National Institutes of Health (NIH) continue to fund studies of biomedical technology and carcinogenic substances.[11]

From a political perspective, several remarkable trends exist in federal-agency acceptance and utilization of technology assessment. In spite of endorsements at high levels of federal agencies, technology assessments apparently were never fully integrated into agency decision-making processes. The DOT program was for a time more influential than the others, in part because it was located in the secretary's office near the center of departmental decision making, whereas other programs are or were in agency R&D divisions. Moreover, in nearly every case in which an agency established a formal technology-assessment program, one primary and notable objective was intradepartmental integration: for example, in the DOT, integration of the planning for several modes of transportation; in the EPA, better integration of R&D activities with the work of the regulatory divisions; in the DOA, integration in the sense of comparison and trade-offs among benefits and costs for agribusiness, small farms, and consumers, the agency's disparate constituencies; and in the DOE, integration of energy and environmental responsibilities.[12] The other science- and technology-related agencies conducted or sponsored technology assessments without establishing a continuing program or an organizational mechanism for undertaking more than ad hoc technology assessments.

In historical perspective, federal-agency conduct of technology assessments reveals several important patterns. Technology assessments were concentrated on areas of intense political concern and conflict rather than directed to the broader range of advancing technologies. The number of technology assessments that was done by or for agencies during the 1970s has been estimated at about fifty.[13] Those studies were primarily on high-visibility energy technologies and problems though transportation, agriculture, and water projects received considerable attention as well. There also appears to have been a tendency to narrow or limit the scope of technology assessments and to address only the immediate concerns of agency heads. Federal agencies rarely undertook open-end searches for long-range, indirect impacts and tended to give minimal attention to areas where no definitive answers or policy proposals appeared forthcoming. Most agency technology assessments were done by contractors, although some were entirely in-house and a few relied on so-called blue-ribbon panels of outside experts. Contracts were preferred because agencies typically lacked inter-disciplinary staff, in-house experience, and familiarity with technology assessments or could not commit enough staff time to such projects. Since technology assessments must be fitted to the respective agency's mission and relevant to its primary constituency, the further tendency appeared for assessments to be inverted. In other words, projects frequently became (sometimes deliberately, sometimes subtly and unconciously) studies of the impact of socioeconomic changes on technology or an agency's clientele instead of an assessment of the effects of developing technological and industrial processes on society. Finally and least surprising, technology assessments were conducted primarily by agencies oriented toward science and technology—especially physical and engineering-related technologies as compared to biological, medical, and social technologies.

The OTA

The initial recognition that technology assessment was to be done for and used by the Congress was the strongest incentive for its adoption by executive agencies and, perhaps, remains the best safeguard for its continued existence at the NSF. It is therefore appropriate here to briefly examine the history of the OTA and its contribution to the development of technology assessment in the national government.

With a staff of over 100, the OTA remains a small research center for Congress and its committee staffs. The OTA nonetheless evolved from a research-contracting unit, as originally conceived, to an in-house-research group, which may eventually become a crucial policy-planning organization for Congress and federal agencies. The evolution of the OTA was driven by both the dynamics of congressional politics and structural factors inherent in the relationship of the OTA to Congress.

In establishing the OTA, Congress, in the Technology Assessment Act of 1972, specified that, "The basic function of the office shall be to provide early indications of the probable beneficial and adverse impacts of the implications of technology and to develop other coordinate information which may assist the Congress."[14] In its first two years, however, the OTA did not do much technology assessment in the sense of researching potential and unexpected technological impacts. Instead, it mainly conducted technical-feasibility studies and economic analyses as well as reviewed and critiqued research plans and budgets of federal agencies. By 1974, the OTA appeared to be becoming a general-purpose, science-oriented think tank whose role would be limited to supplying Congress with reports on technological developments and their direct economic costs, benefits, and risks. In the mid-1970s, the OTA's advisory council and the larger, mostly academic, technology-assessment community complained that the OTA was doing nothing they could recognize as technology assessment. This criticism had considerable validity. It also registered the fact that the OTA was under extraordinary pressure from Congress to produce a large number of reports on a broad range of technical and scientific matters—matters already subject to federal-agency programs but about which Congress demanded further information.

Initially, the OTA had little internal structure. The first projects were generated almost haphazardly from requests by congressional committees (largely those chaired by powerful members of OTA's governing board).[15] Little or no attempt was made to convert these numerous requests, framed around the need for technical information, into broader impact assessments. This apparently was a deliberate strategy adopted by the OTA's first director, ex-Congressman Daddario. He sought to build a congressional constituency through establishing a record of responsiveness to congressional requests. Some of the early studies continued over several years and gave rise to committee requests for further studies and, ultimately, led to the establishment of continuing programs on, for example, oceans, energy, transportation, food, and materials. Thus, OTA staff came to concentrate on particular technology areas and to establish important links with congressional subcommittees, thereby potentially influencing the direction of legislative policy. Indeed, a series of in-depth interviews with OTA program managers revealed that they had come, by the mid-1970s, to believe that they understood congressional concerns and directions better than committee staffs, who were usually not scientifically trained.[16] OTA program managers said they were steadily moving away from technical advice and toward policy analysis and planning. As one program manager insisted, "Congressional committee staffs are not able even to recognize policy issues [related to science and complex technology] and that OTA's most necessary function is to identify, define, and analyze science and related policy issues for Congressmen."[17]

The OTA's second director, Russell Peterson, took office in 1976 after Daddario resigned. Peterson was no less committed to the enterprise of technology assessment.[18] To the contrary, he wanted the OTA to move more positively and aggressively in fulfilling the early-warning and social-impact-assessment functions mandated by Congress. For Peterson and some OTA staff, this implied policy planning—the OTA should identify assessment needs, explore emerging technologies, and bring those issues and reports to the attention of Congress. Those who define the issues, of course, take the first step in formulating public policy. When creating the OTA, Congress neither intended not anticipated that it would assume a role in stimulating and guiding policy formulation rather than passively supporting congressional decision making with technical information. Yet in 1977, program managers and their staff believed that they were so deeply immersed in research of national problems that they could anticipate emerging policy issues well before those issues become critical enough to force congressional attention. The OTA's relationship with Congress showed signs of shifting from the formal leadership of the Technology Assessment Board to internal and informal sources of leadership within the OTA as program managers developed closer connections with congressional staff. In sum, in 1977 the OTA appeared on the threshold of becoming a policy-planning center, oriented toward subtly directing congressional attention to science- and technology-policy issues.

Whether or not objections of some congressmen to the apparently emergent activism at the OTA led to Peterson's abrupt departure, after only nine months in office, remains a matter of conjecture.[19] In any event, the third director, John Gibbons, formerly of the Oak Ridge National Laboratory, moved to placate the OTA's congressional board. He returned the office to a low-visibility profile, held down the size of the staff, completed all ongoing studies, and initiated only tightly bound studies in direct response to congressional requests. Still, at the same time and in response to staff requests, he did approve a number of small, in-house background, or planning, studies to explore emerging technologies.

In less than a decade, the OTA published over 100 reports. Although some of the reports cannot be labeled technology assessments as such, many others are very good examples of technology assessment with broad coverage of the social and environmental impact of developing technologies and systematic attention to alternative technologies and policies. OTA studies typically give considerable attention to policy issues and usually set out a series of policy options or alternatives. The OTA does not (and in fact is forbidden) to make explicit recommendations to Congress. The orientation toward specific policy issues—defined in the context of congressional interests, responsibilities, and powers—largely accounts for the differences between OTA reports and those produced by federal agencies. This also explains why the OTA emerged not as a research-funding organization, as originally envisioned, but as an organization producing its own reports and relying on contractors for chiefly data-gathering purposes. The OTA has completed reports in a number of areas on a broad range of topics.[20] A partial list of these reports includes the following research areas and topics:

Energy: Shale, oil recovery, coal-slurry pipelines, solar conversion;

Oceans and related technology: Offshore oil and gas, renewable resources, the 200-mile limit;

Transportation: Railroad safety, the automobile, mass transit, community planning;

Communications and computers: Information systems related to the IRS, criminal justice, and banking; rural communications;

Food: Information systems, grading and marketing systems, livestock feed, contaminants, pest management;

Materials: Public-resource management, stock piling, information systems, conservation and recycling, implications of scarcity;

Health: Drug bioequivalence, vaccines, computers and medical information, carcinogenic substances;

Military: MX missiles, implications of nuclear war.

The major problem with OTA assessments is that they tend to run for so long (sometimes two to three years) that congressional interest wanes or Congress is forced to act before the study appears. Because the OTA undertakes studies at the request of congressional committees, moreover, congressional members already recognize the importance of the policy issue or potential problem, and hence they rather than the OTA provide the early warning of emergent policy issues. For these reasons, Director Peterson and many observers outside of Congress and the OTA continue to insist that the OTA initiate some studies on its own.

According to both OTA program managers and congressional staff, OTA reports and background documents do appear to be used by committees preparing for hearings and in considering and drafting legislation. OTA reports are cited in congressional debates, often by both sides, which is not a measure of ambiguity of the reports, rather an indication that the OTA is fulfilling its primary function of supplying balanced, usable information on science and technology issues to the Congress. The OTA remains a unique mechanism for generating scientific and technological information and thereby informing congressional decision making. It has the potential of becoming a crucial planning vehicle for Congress and thus an even more-remarkable innovation in democratic governance.

A Retrospective and Prospective View

Technology assessment fundamentally rests in the assumption that the government, in a pluralistic and heterogeneous society, provides obstacles and opportunities for balancing and resolving competing values and contending interests. In the 1970s, technology assessment emerged as a discrete, coherent category of public-policy analysis, conducted by various institutions within the national government, and aimed at enhancing public management and policy under conditions of uncertainty and intense value conflict. Technology assessment basically represents a philosophy of decision making, an attempt to build into national political institutions and processes mechanisms for identifying in a balanced fashion the unexpected, unplanned, and undesirable effects of advancing technologies. Although technology assessment involves interdisciplinary social-science research, it remains vulnerable both to liberal critics who allege that it has a protechnology/proengineering and antidemocratic bias and to conservative critics who perceive it as antitechnological, antimarket, and a pretense for governmental intervention into the private sector. In the 1980s there indeed appears to be a backlash against applied social-science research in government, in part because it was identified in the late 1960s and throughout the 1970s with liberal policies and causes such as civil rights, equality of opportunity, the war on poverty, and environmental conservation. Although the future of technology assessment as a discrete activity within the national-governmental process appears uncertain, its underlying philosophical perspective may well have a more-enduring and profound influence on politics in the United States.

Notes

1. The term *policy analysis* is usually used in the literature of political science and public administration to refer to evaluations of public policies; whereas practitioners typically use the term to refer to the development of an information base and analytical framework in support of policy formulations. I use the term in the latter sense primarily, although the two usages are compatible.

2. Throughout the chapter the author makes explicit judgments as to what is and is not technology assessment. Those judgments are based on the definition and discussion of technology assessment in this paragraph. Much of the material incorporated into this chapter is drawn from two earlier works, both of which reported on studies supported by the National Science Foundation: Vary T. Coates, *Technology and Public Policy: The Process of Technology Assessment in the Federal Government,* NTIS #PB–211453/AS (A15), #PB–211454/AS (A12), and #PB–211455/AS (A03) (Washington, D.C.: George Washington University Program of Policy Studies in Science and Technology, July 1972); and Coates, *Technology Assessment in Federal Agencies, 1971–1976,* NTIS #PB–295969/OSL (A25) (Washington, D.C.: George Washington University Program of Policy Studies in Science and Technology, March 1979). These two works are hereafter cited as Coates (1972) and Coates (1979).

3. See, for example, Resources for the Future's national public-opinion survey in 1980, conducted for the Council on Environmental Quality, in *Environmental Quality,* Eleventh Annual Report of the Council on Environmental Quality (Washington, D.C.: Government Printing Office, 1980), pp. 401–425.

4. See, for example, U.S. Congress, House, Committee on Science and Astronautics, Subcommittee on Science, Research, and Development, *Proceedings,* 90th Cong., 1st sess., 1967; and Emilio Q. Daddario, "Technology Assessment Legislation," *Harvard Journal on Legislation* 7 (1970): 507.

5. Typical examples that were widely publicized include the Donora air-pollution disaster in 1948; the cranberry-bog scare and the diethylstilbestrole in the chicken-feed issue in the 1950s; and in the 1960s, the thalidomide tragedy, X-radiation from color TV, lampreys in the Great Lakes, asbestos fibers, mercury residues in fish, the Torrey Canyon and Santa Barbara oil spills, foaming detergents in streams, the Dugway sheep kill, and DDT. See, generally, Edward W. Lawless, *Technology and Social Shock* (New Brunswick, N.J.: Rutgers University Press, 1977).

6. National Environmental Policy Act of 1969, 42 U.S.C. §§4321–4361.

7. For a further discussion, see Coates (1972).

8. For example, the OMB reportedly struck funds for impact assessments from the budget of the Federal Energy Agency in 1976 on the grounds that it was intended to be an action agency. See Coates (1979).

9. Such federal-agency studies would include, for example, the DOI's Alaska Ramparts Dam Study, the NIH's Cardiac Replacement Study, the Forest Service's Forest Management Study, and the DOT's Northeast Corridor Project—all projects completed between 1969 and 1971.

10. See Coates (1972).

11. For a further discussion, see U.S. Congress, Office of Technology Assessment, *Assessment of Technologies for Determining Cancer Risks from the Environment* (Washington, D.C.: Government Printing Office, 1981).

12. The DOI, which also had conflicting or competing objectives, developed a mechanism called Program Decision Option Documents (PDOD) about 1975 to inform and support final resolution of interdepartmental policy options at the secretary's level. PDODs focused on specific, immediate decisions but had many of the characteristics of technology assessments.

13. See Coates (1979).

14. Technology Assessment Act of 1972, Pub. #L. 92-484, 85 Stat. 797 (13 October 1972).

15. For example, two of the earliest reports were on drug bioequivalence and automobile-collision data.

16. These observations are based on a study conducted for the NSF. See Coates (1979).

17. Ibid.

18. These observations are based on interviews conducted during 1977–1978.

19. Peterson also took the decisive steps to remove political appointees from the office and to strengthen the prerogatives of the director vis-à-vis the board. He resigned to head the National Audubon Society.

20. A full list of reports prepared by the OTA is available from the OTA's Publication Office, 600 Pennsylvania Avenue, S.E., Washington, D.C. 20003.

Congressional Use of Technology Assessment

David Whiteman

Congress and the OTA

The congressional OTA operates within a difficult organizational environment—one that is sometimes hostile, sometimes supportive, but largely indifferent to its work. Congressional politics has traditionally not been characterized by an interest in comprehensive, long-term assessments of the "physical, biological, economic, social, and political effects" of technological applications.[1] For those interested in the actual and potential impact of technology assessment, however, the establishment of the OTA, a unique organizational arrangement, provides an opportunity to study the use of technology assessment in the complex and dynamic process of public-policy formulation and its impact on congressional politics.[2]

Congressional use of OTA reports is only one of four organizational goals of the OTA. The first and most immediate goal, which drives the day-to-day activities of OTA staff, is the development and completion of high-quality analytic reports. Maximization of congressional use of OTA reports in policymaking is a second goal, most prominent during the beginning and end of project development. A third goal is the translation of congressional use of reports into increased organizational prestige. Attainment of these three goals supports the final and most general organizational goal: maintenance and expansion of congressional support for the survival and enhancement of the OTA. In other words, attainment of the OTA's institutional goal—congressional support—depends on the continued production of projects (its major organizational output), the use of those projects in Congress, and the acquisition of prestige derived from that use.

This chapter focuses on and examines the two intermediary goals of the OTA. Ostensibly, the two organizational goals have a fairly straightforward relationship—namely, the greater the congressional use of OTA projects, the greater the organizational prestige of the OTA. However, a more-careful, detailed analysis, distinguishing between two broad categories of use (strategic and substantive) and among three components of prestige (visibility, nonpartisanship, and impact) reveals an interesting paradox: The types of congressional use of OTA reports that promote visibility are unlikely to

Research for this chapter was supported by the NSF, grant number SES-8009437.

promote perceptions of the OTA's nonpartisanship and impact on the legislative process.

At the outset, one limitation of this study should be noted. The relationship between congressional use of OTA reports and OTA prestige, examined here, is confined to interactions between Congress and the OTA. This focus ignores both the sometimes extensive use of OTA projects by the executive branch, state governments, academia, and numerous private organizations and the prestige the OTA derives from such external and broader uses of its reports. While these external routes to prestige may indirectly promote congressional support for the survival and enhancement of the OTA, the critical link examined here is between the OTA and its primary constituents in Congress.

This chapter draws upon interviews and document analysis conducted in 1979 and 1980 as part of a larger study of the use of OTA projects by congressional committees.[3] That larger study focused on the use of five OTA projects and was based on extensive open-ended interviews with the fifteen OTA staff members responsible for the projects and with thirty-two committee staff members representing the twenty-five different congressional committees that received the completed OTA projects.[4] These interviews, along with content analysis of all formal congressional documents relevant to each project, were analyzed and compared in order to determine the range of OTA project use within those committees. In assessing congressional reliance on the OTA, an OTA project was defined to include the communication of findings through both formal channels such as the actual OTA report and informal channels such as personal consultations between OTA and committee staff.

Strategic and Substantive Use of OTA Reports

Analysis of the paradoxical relationship between maximizing congressional use of OTA reports and maximizing the organizational prestige of the OTA requires, initially, consideration of the meaning of use within the congressional context. Rather than begin with a preconceived definition of congressional use, my investigation left the definition open to the participants in the policymaking process. Committee staff reported use of OTA projects in a wide variety of activities, ranging from planning committee agendas, to clarifying technical details in proposed legislation, to writing floor speeches for members of Congress. Congressional use of OTA reports thus appears to be multifaceted. However, all congressional uses of OTA reports appear to be ultimately instrumental. Not surprising, members and staff of congressional committees appear to have little time or opportunity to use OTA reports for intrinsic purposes—for example, cultivation of knowledge for knowledge's sake. Whatever value OTA reports have for Congress, it lies in their utility for formulating public policies and, more generally, informing the legislative process.

Within a broad characterization of congressional use of OTA reports as instrumental, it is still possible to distinguish between two general categories of such use: (1) strategic and (2) substantive. To illustrate these two different types of use, two case studies are presented. The OTA reports examined here, on coal-slurry pipelines and residential-energy conservation, were the two most widely used of the five OTA projects studied. Yet the patterns of congressional use differed markedly. Congressional use of the coal-slurry report can be characterized as strategic—that is, the substance of the OTA project was less important than the support it offered to partisans in congressional deliberations. Congressional use of the residential-conservation project can be characterized as substantive—that is, the findings of the OTA project appeared in fact to influence the actual content of legislative proposals and to prove instrumental in the formulation of public policy.

The Coal-Slurry Pipeline: Strategic Use

The OTA's coal-slurry-pipeline project, completed in March 1978, addressed what had become a highly salient and controversial topic in Congress: the desirability of seemingly innocuous pipelines to carry coal, in slurry form, from mines to utilities.[5] The two central issues in the controversy involved the possible effects of the pipeline on the railroad industry and on the water resources in western states. Among the secondary issues, difficult enough in their own right, were the potential enviornmental effects of pipeline construction and transportation-regulation questions.

The particular issue confronting Congress was whether the right of eminent domain should be granted to the pipeline corporations in order to facilitate the construction of the coal-slurry pipelines. This aspect was especially controversial since the railroads were reluctant to grant a right-of-way to a competing form of transportation. A coal-slurry bill passed in the Senate in 1974 but died in the House Interior Committee when railroad management, railroad labor, and environmentalists lined up against the construction corporations, utility companies, and energy-development groups. In the Ninety-fourth and Ninety-fifth Congresses (1975–1976 and 1977–1978), the Senate was unwilling to reconsider the issue until the House took action. As a result, most of the activity subsequent to completion of the OTA project in January 1978 took place in the House.

The OTA project attempted to assess a wide range of the direct and indirect implications of pipeline construction. For instance, it compared the costs of transporting coal by pipeline with the costs of railroad transportation, under several different scenarios. In some circumstances (assuming no change in the regulatory environment), pipelines were found to be more economical. The OTA also addressed potential pipeline impact on the railroad industry, on water resources, and on the environment and attempted to summarize the relevant laws and regulations bearing on the general issue. The OTA, however, offered no specific policy recommendations.

Five congressional committees became involved in coal-slurry-pipeline deliberations in the Ninety-fifth Congress. The following list summarizes the legislative activities and documents related to this controversy:

House Commerce Committee:
Request, October 1975;
Hearings, February 1978;
Committee print, June 1978;

House Interior Committee:
Hearings, January 1978;
Markup, February 1978;
Report, March 1978;

House Public Works Committee;
Hearings, April 1978;
Markup, May 1978;
Report, May 1978;
Floor debate, July 1978.

Senate Commerce Committee:
Request, February 1976;

Senate Energy Committee:
Request, May 1976;
Hearings, June 1978.

The Senate committees, awaiting House action, played only a very minor role. In the house, the coal-slurry bill was referred jointly to the Committees on Interior and Public Works. These committees were favorably disposed to construction of the pipelines, although significant dissent remained over concerns about possible depletion of water resources. By contrast, the House Commerce Committee, denied jurisdiction on the bill, attempted throughout the deliberations to gain jurisdiction in order to express its strong opposition to pipelines based on its view that the pipelines posed a threat to the economic well-being of the railroads.

In the spring of 1978, the Interior Committee (by a vote of 30 to 13) and the Public Works Committee (by a vote of 23 to 20) reported the bill to the full House. The Commerce Committee failed in its attempt to gain jurisdiction. However, the water-related concerns of congressional members from western states, combined with persistent lobbying of representatives from the northeastern states by the railroads, resulted in defeat of the legislation on the House floor. The final vote, in July 1978, was 246 to 161.

Congressional reliance on the OTA actually began before the office commenced its coal-pipeline project. The House Commerce Committee's original request for the project cited the need for more knowledge in order to understand "the impact of a coal slurry pipeline on three major areas: (1) energy, (2) the environment, and (3) the economy, including the economic effects on the railroads and the public."[6] Proponents of coal-slurry legislation claimed that the request of a lengthy study of pipelines was politically motivated, designed to end consideration of the legislation in the Ninety-fourth Congress and to delay consideration in the Ninety-fifth. Staff of the House Commerce Committee and the Senate Commerce Committee, which also requested the study in May 1976, did not deny such motivations.[7]

The controversy emerged fully in the summer of 1976 when the coal-slurry bill was awaiting final markup in the House Interior Committee. Opponents of the bill encouraged the OTA to approve the study request. The OTA, however, decided to defer any action on the request until the Interior Committee resolved its position on the bill. When the committee eventually tabled the bill, the OTA then moved to approve the request in July 1976.

The controversy mounted while the OTA developed its coal-slurry project. In the Ninety-fifth Congress, those opposed to the pipeline legislation contended that any action on the bill should await completion of the OTA project. Proponents disagreed, arguing that the study was holding up the work of Congress. Representative Morris Udall, chairman of the House Interior Committee, went so far as to offer $100 to anyone who could show afterward that an amendment to the bill had come out differently because of the OTA study. Eventually Representative Udall, placed in an awkward position as a coal-slurry proponent and also a long-time OTA supporter, reluctantly agreed to postpone final markup of the bill until release of the OTA report.

The preliminary draft of the OTA project became available in January 1978 at the beginning of the second session of the Ninety-fifth Congress. Until then, the only congressional use of the OTA's project was limited to attempts by opponents of coal-slurry pipelines to manipulate the agenda of congressional committees. After release of the report, however, two additional kinds of congressional use of the report became available: the support and documentation of existing positions and the clarification of technical legislative details. Reliance on the OTA project in the House Commerce and Interior Committees illustrates these additional types of congressional use of OTA reports.

The most extensive congressional use of OTA's coal-slurry project occurred in the House Commerce Committee. The committee had a strong, long-standing, and well-formulated position on the pending legislation. As expressed by one staff member: "We were fighting for jurisdiction. We were fighting the whole concept of coal slurry pipelines. We were fighting the bill, in particular." The House Commerce Committee used the project in two main ways: first, to obtain jurisdiction on the bill and, second, to put the OTA report under cross-examination in order to balance out the Interior Committee's portrayal of the report as supporting the propipeline interests. Committee staff reportedly used OTA documents in formulating questions for the hearings held on the bill. At the hearings, both proponents and opponents of the bill couched arguments in terms of the report's findings. Hence, what the report really said became a major contested issue in the hearings. The report basically served as a baseline and focal point for testimony—a common ground from which all could make arguments.

In the larger congressional debate, committee staff used the report both because it was good and for political reasons. The staff reportedly perceived that pipeline proponents had oversold the support that the OTA report gave their position, and they therefore made a "conscious decision" to "match them punch for punch." The OTA report thus appeared frequently in advocacy material (drafts of speeches and other reports and committee prints) produced by the staff.

The House Interior Committee also brought strong policy predispositions—both for and against pipeline legislation—to bear on its use of OTA material. As in the Commerce Committee, staff used the OTA report primarily in supporting and documenting positions in congressional debate. The committee report produced by the staff even reprinted six pages of the OTA report in its entirety in addition to frequently citing the document. The staff likewise used the report in writing issue summaries and speeches for members interested in taking an active role in the floor debate.

Because the Interior Committee had jurisdiction over the bill, the report proved especially useful in its attempts to modify the bill. One amendment, which OTA conclusions were used to support, had the effect of enlarging the number of executive agencies involved in the pipeline-approval process. Another amendment was offered by a member who explicitly "tied the rationale for his amendment to the OTA report," even though his long-standing position on the amendment predated the report. A third amendment apparently stemmed more directly from the OTA project. The OTA report indicated potential problems with the bill and specifically that its language might "leave little scope for state regulation of water for a coal slurry pipeline."[8] In response, committee staff wrote "some stronger language" and "cleaned up some of the language to take care of the . . . technical concerns that OTA had." This technical clarification benefited those members concerned about state water rights, although the extent to which they advocated the change remains unclear.

Congressional use of the OTA's coal-slurry-pipeline report illustrates the varieties of strategic use of OTA reports. The report was used in attempts to manipulate the agenda, to support policy positions, and to support modifications in pending legislation.

Residential-Energy-Conservation Legislation: Substantive Use

Energy was among the most salient political issues in the late 1970s. Responding to a request from the Senate Commerce Committee, the OTA commissioned a study of residential-energy conservation in 1976. After a long period of project development, the OTA finally delivered its report to Congress in July 1979.[9] Release of the report came just in time for the debates over several energy-conservation initiatives in the Ninety-sixth Congress.

The central focus of this controversy was the Energy Security Act, better known as the Synfuels Bill.[10] After synfuels legislation was proposed in the summer of 1979, members of both houses, but particularly the Senate, began to argue that inclusion of other energy-development programs in the bill was a necessary precondition for their support of the legislation. This ultimately resulted in the adoption of several additional titles in the bill, including provisions for biomass and alcohol fuels, renewable energy sources, solar energy and energy conservation, and geothermal energy. The conservation section of the revised bill provided for a solar-energy and energy-conservation bank, a residential-energy-efficiency program, a weatherization program, and programs for energy-auditor training and commercial and industrial conservation. The act passed the House (by a vote of 317 to 93) and the Senate (by a vote of 78 to 12) in June 1980. In addition to the Synfuels Bill, two other legislative concerns about residential-energy-conservation issues included a building-energy performance-standards program and a community-energy-efficiency program. The committees involved in all these activities and the documents produced are listed in the following list:

House Banking Committee:
Hearings, October 1979;
Report, November 1979;

House Commerce Committee:
Hearings, September and
October 1979;
Report, December 1979;

House Government Committee:
Hearings, July 1979;
Report, November 1979;

Senate Banking Committee:
Hearings, July, October, and
December 1979;
Report, October 1979;

Senate Commerce Committee:
Request, February 1976;

Senate Energy Committee:
Hearings, March, April, July,
and December 1979;
Report, October 1979;

House Interior Committee: Senate Government Committee:
Hearings, July 1979;   Hearings, November 1979;

House Science Committee;   Senate Floor debate, June and
                                   November 1979 and June 1980.
House Task Force on Energy;

House Floor debate, June 1979   Joint Economic Committee,
and June 1980.
                                   Joint Conference Committee:
                                   Report, June 1980.

The OTA report, *Residential Energy Conservation,* attempted to provide a comprehensive assessment of the potential for residential-energy conservation. After estimating future trends in residential-energy consumption, the project considered the role of major political actors and institutions including consumers, builders, utilities, oil companies, local governments, state governments, and the federal government. The OTA also summarized information about some of the new technologies available for conservation in the residential sector. As required by Congress, the report made no specific policy recommendations.

The wide range of congressional use of the OTA report on residential conservation is perhaps best illustrated by focusing on three of the eleven committees that used the report: the House Commerce Committee, the Senate Energy Committee, and the Joint Economic Committee.

The Joint Economic Committee's use of the OTA report demonstrates the report's strategic value in congressional politics. Two committee staff members had developed an energy-conservation bill during the summer of 1979, at the request of Senator Edward Kennedy. Because the Joint Economic Committee lacked legislative jurisdiction over the area, the bill was eventually sent to the Senate Energy Committee for consideration. During subsequent negotiations with the Energy Committee, staff of the Joint Economic Committee approached the OTA project with very narrow purposes—namely, its use in their advocacy of their bill. More specifically, the part of the OTA project used was the estimation of potential savings from residential conservation. The staff felt that the report made "a very comprehensive, thorough argument for the position that there is enormous potential" for energy savings. As committee staff observed: "We weren't asking them for political advice. We were just asking them for expert testimony . . . just the numbers, and how they arrived at them." The use of the OTA's estimations of energy savings in congressional advocacy was actually twofold. In deliberations within the Senate Energy Committee, staff of the Joint Economic Committee used OTA estimations of costs and benefits specifically and directly to "bolster [their] argument[s]," even to the point of simply copying pages of the OTA report. In Senate deliberations, staff "used the numbers that are in the [OTA] report as the basis for speeches, statements, and fact sheets that accompanied [Senator] Kennedy's bill."

Substantive use of OTA projects remains illustrated by the activities of the House Commerce and Senate Energy Committees in considering residential-energy-conservation legislation. Use of the OTA report by the House committee centered primarily around its drafting of a conservation bill. Within the process of formulating legislation, a constant dialogue took place with OTA staff, who were involved even in the earliest brainstorming sessions. OTA staff helped in defining the problems to be addressed in legislation. Congressional staff said that the OTA report was very useful in "framing not just [the proposed] legislation but general policy statements and ideas for how to get things done." At a more-technical level, the OTA provided information "on numbers of houses and levels of insulation" involved in various proposals for residential-energy conservation and reviewed drafts of proposed legislation. In general, House Commerce Committee staff found it difficult to "give OTA credit specifically" for any particular aspect of the legislation but then again acknowledged that the OTA was "very much involved" in its formulation of policy objectives and proposed legislation.

The Senate Energy Committee made the most varied use of OTA material of all the committees investigated. The Senate committee primarily used OTA reports prior to passage of the Synfuels Bill. While the orientation of the committee toward residential conservation had been formulated without OTA involvement, some provisions, as well as later deliberation and bill markups, clearly exhibited OTA influence. Indeed, for conservation-related provisions, staff claimed that the OTA report "was one of the primary documents that [they] used." OTA information also helped clarify more-technical aspects of the provisions. For example, one part of the legislation, modeled after a Canadian program, had to be adapted for application to U.S. conditions. Committee staff determined how many people lived in different parts of the country, in different economic ranges, and then consulted the OTA to "figure out how much conservation was possible under different scenarios, and so on."

During committee markup of the bill, the OTA again served to clarify the costs and benefits of the proposed legislation. As a committee staff member observed:

> As the committee progressed through the development of that . . . program, kept modifying it and amending it, a lot of the information that we dug our fingers into, that helped us modify in the right ways to keep it intact and not have the committee completely emasculate it, came from the OTA report.

One frequent question brought up during markup of the bill concerned the amount of oil expected to be saved as the result of various conservation measures. Committee staff relied on the figures of the OTA report to calculate these estimates. Twice during markup, committee staff explicitly mentioned the OTA report in answering questions from senators.

The OTA report was further used in other committee activities, although to a lesser extent. At the committee hearing, the report was cited six times. In addition, several interest groups that had provided information to the committee cited the OTA report as backup. During consideration of the legislation on the Senate floor, the report was used "as a tool to convince Senators, and then for the Senators to convince other Senators, that this is a credible thing that they're doing." Committee staff supported these efforts by providing information that "a lot of the memorandums that we wrote to the Senators who were backing the conservation part of the Energy Security Act came from [the OTA] report." The report also was cited in press releases and speeches to bolster conservation provisions.

Despite these occasional strategic uses of the OTA residential-energy-conservation report, overall congressional use of the report seems primarily substantive. The diffuse nature of legislative formulation, with multiple sources of information bearing on any one point, prevents precise identification of the specific changes that stemmed solely from substantive use of the OTA findings. However, based on perceptions of committee staff involved, substantive use of the OTA report was made in defining problems, suggesting approaches to solving problems, and clarifying the technical basis of proposed solutions.

The OTA, Institutional Prestige, and Congressional Politics

Both the residential-energy-conservation and coal-pipeline reports were used extensively by the members and staff of congressional committees. The nature of congressional use of OTA reports varied considerably. The residential-energy-conservation report was used in drafting the substance of legislation, while the coal-pipeline report was primarily put to strategic use in the process and politics of policy formulation. This variation, in turn, has important implications for the ability of the OTA to translate congressional use of its reports into increased organizational prestige.

Within the congressional environment, three components of the OTA's institutional prestige are particularly important: its visibility, perceived nonpartisanship, and perceived impact. The differences between the contributions of substantive and strategic congressional use of OTA reports to the overall organizational prestige of the office can be assessed according to these three components.

Institutional Visibility

Unlike other congressional-support agencies such as the Congressional Research Service and the General Accounting Office (GAO), the OTA does not produce projects specifically for individual members of Congress. One consequence of this is that the OTA must more heavily depend for its visibility on avenues other than direct contact.

The number of citations of OTA reports in congressional documents is one indicator of the OTA's visibility within Congress, and here the difference between the coal-pipeline and residential-energy projects is quite dramatic. According to table 4-1, the coal-pipeline project was cited 616 times in committee hearings, 20 times in committee markup sessions, 32 times in committee reports, and 41 times in floor speeches. The high degree of conflict surrounding the issue, and the ability of both proponents and opponents to use the OTA report, promoted extensive and visible use of the report. In contrast, the residential-energy-conservation report was rarely cited: 11 times in committee hearings, 2 times in committee mark-up sessions, 20 times in committee reports, and not once in floor speeches. Thus, use of the residential-energy-conservation report was extremely difficult to document. The conclusions of the OTA report were communicated to congressional members and staff via occasional briefings and consultations with OTA staff. However, the energy-conservation components of the Synfuels Bill were, in relation to the other components, less controversial, and the OTA report never became a weapon in the arsenals of opposing interest groups and factions within and without Congress.

In general, substantive use of OTA reports in Congress appears to be much less visible and much more difficult to document than strategic use. Although use of both the coal-slurry-pipeline and residential-energy-con-

Table 4-1
Citations of OTA Reports as an Indicator of Visibility

Source	Coal-Slurry Pipelines	Residential Energy Conservation
Committee hearings	616	11
Committee prints	52	0
Committee markups	20	2
Committee reports	32	20
Congressional Record	41	0
Total	761	33

servation projects was extensive, congressional members and staff were much more aware of the OTA coal-slurry-pipeline report because of the nature of the controversy and due to the findings of the OTA. In other words, issues characterized by high conflict and high salience, like coal-slurry pipelines, give rise to greater opportunities for extensive and strategic use of OTA reports.

Nonpartisanship

Although high-conflict policy issues provide greater possibilities for OTA's visibility, they also carry a greater risk. Crucial to the OTA's organizational prestige, and ultimately to its survival, is the maintenance of a nonpartisan image—both in any given policy debate and in interparty struggle. (The 1981 change in party control of the Senate reinforces the import of this latter point.)

While every project carries with it some degree of political risk, some possibility of alienating certain interests and members aligned with those interests, the greater the conflict within Congress over the issue, the greater the risk. In this respect, the OTA's decision to do a coal-slurry-pipeline study was much riskier than its decision to do a residential-energy-conservation study because involvement by the OTA in high-conflict policy issues is always potentially dangerous. Moreover, it should be noted that it would have been almost as risky for the OTA to refuse to do the coal-pipeline study.

As it turned out, both sides in the coal-slurry-pipeline debate were able to use the OTA's findings effectively, and the potential damage to congressional perceptions of the OTA's nonpartisanship and prestige was largely avoided. High-conflict policy issues, as evident in the initial stages of the pipeline project, are nevertheless difficult for the OTA to manage. If a project (or even proposed project) were to come to be perceived as an important weapon for one side of a congressional debate, those holding opposing views would be likely to attempt to discredit their sources of information, including the OTA.

Impact

Perception of the impact is probably the most important and most elusive component of organizational prestige. All evaluations of the OTA ultimately address the question: Does OTA analysis make a difference? An adequate response to the question of impact, however, presents numerous problems. For example, documentation of the impact of policy analysis in any context has proved difficult, and these difficulties are exacerbated by the multiplicity of policy influences present in the congressional environment.

Strategic use of OTA reports by congressional members and staff seems to violate traditional conceptions of the use of policy analysis. OTA reports are often put to strategic use in Congress to reinforce congressional positions taken long before dissemination of any OTA research findings. The impact of congressional strategic use of OTA reports therefore must focus not on the alteration of congressional positions but on the potential increase in the clarity and sophistication of congressional debates. Even though such impact is not essentially documented, it seems clear that if the OTA is perceived as only reinforcing existing positions, and not contributing anything to the legislative process, then its organizational prestige suffers.

Substantive use of OTA reports in Congress conforms to traditional conceptions of the role of analysis in policymaking: The findings of analysis are used to alter the substance of positions or proposed legislation. Yet, as the residential-energy-conservation project indicates, Congress's substantive use of the OTA reports occurs within a complex web of sources consulted by congressional members and staff. Attributing any single legislative provision to any single source is extremely difficult. Thus, the OTA can rarely point to specific substantive applications of its projects and to the direct, tangible impact of its reports.

Conclusion

The translation of congressional use of OTA reports into organizational prestige for the OTA does not happen directly or automatically. The two broad categories of congressional use of OTA reports distinguished here—substantive and strategic—differ greatly in their potential for enhancing the OTA's visibility, nonpartisanship, and impact on the legislative process. Although substantive use carries little risk of partisanship and conforms to traditional conceptions of impact, it does appear to enhance the OTA's visibility within the congressional environment. Strategic use of OTA reports, while very visible, bodes the risk of transforming the OTA into a partisan organization within congressional politics.

The crucial process of translating congressional use of OTA reports into institutional prestige increases in importance to the extent that the OTA exercises control over the subject matter of its projects. Usually the OTA can influence the type of use made by its reports and, therefore, can gain some control over the potential prestige generated by its reports by assessing the amount of political conflict attending the policy issues it accepts to study. In sum, high-conflict environments will yield visibility for the OTA but also will pose certain political risks. Low-conflict environments entail little risk for the OTA's institutional prestige but provide little visibility. Neither type of policy environment is likely to promote congressional perceptions of the overall impact of the OTA on the legislative process.

Notes

1. Pub. L. 92–484, Technology Assessment Act of 1972, §2(d)(1).

2. Congressional use of policy analysis, in any form, has rarely been studied empirically. The two studies most relevant to the research reported in this article are Margaret Boeckmann, "Policy Impacts of the New Jersey Income Maintenance Experiment," *Policy Sciences* 7 (1976):53–77; and Mark Berg, Jeffrey Brudney, Theodore Fuller, Donald Michael, and Beverly Roth, *Factors Affecting Utilization of Technology Assessment Studies in Policy-Making* (Ann Arbor: Institute of Social Research, 1978).

3. See David Whiteman, *Congressional Use of Analytic Knowledge* (Ph.D. diss., University of North Carolina, 1981).

4. All quotations used in the text are from congressional committee staff, who were guaranteed anonymity. It should be noted that the decision to interview committee staff after most potential project use was completed, as opposed to interviewing them throughout their use of the project, necessarily introduced problems of recall. Two techniques limited the effects of this problem. First, a detailed chronology of each project was reconstructed from OTA and committee files and refined throughout the interviewing process in order to remind respondents of specific events and the context of those events. Second, the open-ended-interview strategy increased the likelihood that respondents would accurately recall and report past events.

5. U.S. Congress, Office of Technology Assessment, *A Technology Assessment of the Coal Slurry Pipelines* (Washington, D.C.: Government Printing Office, 1978).

6. Letter of request to the OTA from Representative Fred Rooney, chairman of the House Commerce Committee's Subcommittee on Transportation and Commerce, 21 October, 1975.

7. Another motivation of the House Commerce Committee was its desire to obtain support for its claim of jurisdiction over the pending legislation.

8. U.S. Congress, *Technology Assessment of Coal Slurry Pipelines,* p. 132.

9. U.S. Congress, Office of Technology Assessment, *Residential Energy Conservation* (Washington, D.C.: Government Printing Office, 1979).

10. Energy Security Act, Pub. L. 96–234.

5

Environmental Protection and Technology Assessment: Reflections on the Role of Strategic Planning at the EPA

Gregory A. Daneke and
James L. Regens

Regulatory Reform and Environmental Protection

It has become almost axiomatic in the chronicles of administrative policy-making that when Congress and/or the president face a seemingly unresolvable problem, a new agency is created. In the late 1960s, the environment had become such a problem. Like many of the regulatory reforms in the late 1960s and early 1970s, the establishment of the Environmental Protection Agency (EPA) was influenced heavily by state-of-the-art notions about administrative policymaking.[1] Advocates of the creation of the EPA were particularly concerned about administrative discretion and sought to limit the agency's discretion by providing clearly stated objectives, specific timetables, and detailed instructions for implementing the EPA's statutes. Their concern stemmed from the traditional capture of regulatory agencies by the very groups whom they were intended to regulate. Although the EPA has been more or less successful in avoiding capture by special-interest groups, it has been less effective in carrying out its mission; the underlying concepts of administrative policymaking of post-New Deal agencies such as the EPA have proved inadequate for the highly complex, scientific issues entailed by environmental protection.

The EPA, unlike other regulatory agencies, was largely successful in avoiding capture, but critics maintain that it has had more-limited success in accomplishing its overall mission of environmental protection.[2] With the various environmental-protection acts of the 1970s, the EPA was given a legislative mandate that included a mix of both very specific and very general objectives ranging from preventing exposure to toxic substances to improving ambient air quality. Because of the EPA's expanding legislative

---

This chapter draws heavily upon Gregory A. Daneke, "The Future of Environmental Protection: Reflections on the Difference Between Planning and Regulating," *Public Administration Review* (1982). The opinions expressed here are not necessarily those of the U.S. Environmental Protection Agency.

mandate, a number of those objectives are administratively difficult; a problem made more complex because EPA's decision-making practices did not always consider the practicality of implementation.[3] Furthermore, many of EPA's statutory mandates often proved neither "technologically feasible" nor "economically feasible."[4]

Even when attainable from an engineering standpoint, accomplishing EPA's objectives in one media (that is, air, water, soil, biota) posed potential environmental problems in another. For example, reducing oxides of sulfur emissions by using scrubbers on power plants produces solid wastes (sludge), yet solid wastes must in turn be disposed of in an environmentally acceptable fashion. Consequently, despite all of the rhetoric about dealing with the environment in a comprehensive multimedia fashion, much of what Congress mandated the EPA to accomplish within a specific time frame was not scientifically and/or technologically possible. As Alfred Marcus observes, "Air pollution standards, based solely on health-and-welfare criteria, were designed to force technological change. But, legal requirements alone will not compel rapid technological innovation. As a result, serious technological obstacles plagued efforts to reduce emissions from major sources such as steel mills, power-generating plants, and motor vehicles."[5] Hastily developed control technologies, such as the catalytic converter, which reduced one form of pollution while exacerbating others, illustrate the problems involved in meeting unrealistic timetables. No less ill-conceived were some of the specific goals established by Congress. Resource economists, for instance, have been quick to point out that EPA's mandate for zero level of effluent discharges under the Federal Water Pollution Control Act might mean more clean water than the nation is willing to pay for given the rapidly diminishing returns on increased cleanliness. Similarly, in a 1976 report, the National Commission on Water Quality asserted that "many of the delays associated with Federal decisions . . . were unavoidable consequences of unrealistically short statutory deadlines."[6] Finally, while technological feasibility is a necessary condition for achieving environmental-protection goals and economic feasibility is a desirable attribute, neither is sufficient to guarantee political feasibility. The latter involves determining trade-offs among benefits and damages and reconciling those tradeoffs across multiple interests. Not surprisingly then, Lester Lave and Gilbert Omenn conclude that the regulatory practices in pursuit of clean air have been a national disgrace and that much more fine-tuning is required.[7]

Rethinking Environmental Protection: The Role of Strategic Planning

After almost a decade of controversy over environmental protection, there appears to be increasing support for a thorough rethinking of both the theory and practice of environmental protection. That is not to deny that the environment remains an important national policy issue and that most Americans still believe that environmental quality should remain a focal point of regulation. Pollster Louis Harris, for example, found "that by a margin of 80 to 17 percent the public nationwide does not want to see any relaxation in existing federal regulation of air pollution."[8] Still, it seems clear that environmental objectives will be increasingly subject to tradeoffs with other social and political objectives, especially those favoring economic productivity.[9] Indeed, if the environment is a valuable resource like capital and labor, then environmental goals should be included in a system of tradeoffs with other goals, especially economic ones.[10] Whether the present approaches to environmental regulation facilitate meaningful tradeoffs is certainly questionable.

Systematic planning offers one way of reducing uncertainty such as the economics of environmental protection. With specific reference to environmental issues and agencies, Guy Benveniste provides insights into how systematic planning can strengthen the regulatory process.[11] He argues that the current regulatory process relies too heavily on formalistic rule-making procedures and the imposition of sanctions. Procedural fairness required at many stages of policy implementation inevitably invites regulatory delay and judicial intervention. According to Benveniste, the success of this approach demands precise mandates, unequivocal scientific evidence in support of rule making, and equitable effects upon all regulated industries or organizations as well as the ability of the regulated to adapt without external assistance. Ideally, state and local agencies have the financial and technical expertise, as well as the will, to carry out enforcement. Moreover, implementation tasks should be routine with minimal discretion so that all relevant factors can be included and all procedures and participatory requirements adhered to.

Because these factors usually are difficult to obtain in practice, the regulatory approach to environmental management usually bogs down. Indeed, environmental protection and regulation seems to be plagued by several other inhibiting factors as well. First, while the public usually values environmental quality, such support commonly is nebulous and disorganized. Second, the scientific basis for environmental decisions is frequently inconclusive and subject to substantial disagreements and uncertainties. Finally, EPA-prescribed technical fixes for securing pollution control may, in some cases, demand technological and industrial adjustments beyond the capabilities of individual firms and/or generate severe economic dislocations.

Strategic-planning processes, including the use of technology assessments, therefore, seem to be better suited to achieving environmental management. As Benveniste argues:

> A planning approach assumes that intervention has to be first concerned
> with what is both politically desirable and technically feasible. . . . The
> problem is defined so that political agreement is reached and a coalition
> is based on recognition of the economic and technical dimensions of
> proposed improvements. Furthermore, the coalition exists because the
> regulator is able to coerce it with both rewards and punishments.[12]

Such an approach has several advantages. First, by requiring fairly systematic analysis, it focuses attention on the adequacy of existing scientific information for rulemaking or regulatory action. Second, the planning approach helps to identify what type of standards (that is, performance, command, and control) should be developed given the relative reliability and availability of scientific evidence on which to base regulatory standards. Finally, because industry wants to maximize certainty and does not want to compete in pollution control, uniform yet defensible government regulatory actions may provide desirable information for the marketplace.

Some type of cooperative public-private-sector planning as an alternative to environmental regulation—based on governmental sanctions—may be the best way to achieve the broad goals of environmental protection and conservation. Richard Brooks, director of the Environmental Law Center at the University of Vermont, further contends that law may establish planning mechanisms as a control on the administrative process in three ways:

> It may seek to define the method which planning must pursue. It may seek
> to define the components of a plan and require the resultant decision to
> implement the plan. Thirdly, it may seek to define the substantive problems, objectives, or standards which planning must pursue within a given
> subject matter area.[13]

In other words, mediation, negotiation, and other forms of conflict resolution might be viewed not only as alternatives to formal adjudicatory processes for environmental regulation but also as mechanisms for incorporating planning and policy analysis in environmental management.[14]

Planning and Incentive Systems

For some time, economists have argued for reliance upon incentive systems instead of direct regulatory standards.[15] Essentially, they argue for reinstatement of clear market signals and the internalization of the costs of externalities, such as pollution, through the imposition of pollution rights and/or charges (also called effluent fees). While a fee or permit system can be viewed as a penalty, it also represents an economic incentive to limit pollution and to develop new control technologies. If applied properly, incentive systems are much more economically efficient (in a Pareto-optimal sense) than regulatory standards—that is, the public gets just about as much environmental quality as it is willing to pay for. No less crucially, the burden of technological innovation is shifted back to the private sector where it appropriately belongs and where the profit motives are likely to enhance the prospects for creative technological solutions to pollution problems.[16]

Although well developed in theory, governmental provision of incentives for the private sector to undertake pollution controls remains problematic. One of the limitations of incentive systems has been their traditional reliance on a regulatory as opposed to a planning approach to environmental protection. Incentive systems recognize the necessity of specifying social values for heretofore unmeasured environmental protection, but without some planning mechanism the process of valuing the social costs and benefits of environmental protection remains vague, subject to perversion by existing political arrangements and alliances.[17] Furthermore, as Garrett Hardin noted in his classic article, "The Tragedy of the Commons," existing sociopolitical institutions provide substantial incentives for environmental degradation.[18]

Thus, the problem of providing incentives for pollution control appears not so much a matter of correcting market failures but rather of establishing mechanisms for collective choice that are significantly different from existing market mechanisms. Traditional regulation has primarily concentrated on imposing newly articulated values upon existing markets. Planning mechanisms, by contrast, focus on developing new procedures for valuing the social costs and benefits of environmental protection. The development of incentive systems, therefore, may provide the critical juncture at which planning and regulating approaches coalesce in furthering public-and-private-sector cooperation in environmental management.

The shift from primary reliance on a procedural/legalistic approach to environmental regulation and to greater reliance upon strategic planning represents a fundamental reorientation in environmental management. Here again, Benveniste is suggestive in arguing that environmental degradation entails a series of "social errors" by both public and private organizations.[19] In addition to evoking ideas such as corporate social responsibility, he argues that traditionally environmental protection has been ex post facto management—for example, aimed at mitigation of and/or compensation for environmental damages. While remedial at first, a planning process is by design ex ante, directing regulatory agencies toward prevention, as opposed to mitigation of environmental pollution. In other words, strategic planning relies more heavily on new concepts and models such as technology assessment and less on older approaches such as command and control in achieving environmental protection.

Strategic Planning: Theory and Practice

Extolling the potential virtues of strategic-planning processes belies the fact that such an approach has been rarely adopted by federal agencies. Most of what passes for strategic planning in the federal government is actually rudimentary tactical analysis—that is, operations modeling and programming.

To elaborate on this distinction, it may be useful to draw on David Wilson's categorization of planning approaches.[20] He identifies the following major models:

Rational-comprehensive model: centralized, comprehensive, mechanistic;

Incremental model: political, marginal shifts;

Mixed scanning: situational;

General systems: integrative, interactive, goal-defining, iterative;

Adaptive-learning: decentralized, participative, humanistic, futures oriented.

Several federal agencies continue to have a preoccupation with the rational-comprehensive approach to planning, in which a consensus on organizational goals is assumed and the principal activity focuses on translating those goals into operational programs and projects.[21] Surprisingly, this model continues to receive support even in the face of incremental realities—that is, fragmented decision structures and political dissonance—that constrain agency plans from being implemented.

In contrast, strategic planning, including technology-assessment techniques, requires a kind of hybrid form of policymaking that combines elements of the systems and adaptive models.[22] Such a hybrid form of strategic planning is becoming popular in schools of planning, as well as increasingly appreciated in the private sector.[23] Although, as Donald Michael documents, considerable resistance exists to strategic-planning approaches in federal agencies, it is not unreasonable to propose that regulatory agencies begin to adopt this type of approach to environmental protection, just as corporations have begun to accept their social and environmental responsibilities.[24]

Initial attempts to establish a strategic-planning approach to environmental protection within the EPA were evident in several large-scale, regionally based technology assessments of energy resources conducted for the Office of Energy, Minerals, and Industry during the mid-1970s.[25] Those technology assessments were designed to inform public- and private-sector decision makers as well as interested citizens about the likely environmental consequences of developing and operating particular technologies. The assessments also aimed at identifying, evaluating, and comparing alternative policies and implementation strategies.

Drawing upon its experiences with such technology assessments, the EPA in 1980 moved toward improving its capabilities in conducting more-comprehensive planning of environmental management. In particular, the EPA established an Office of Strategic Assessment and Special Studies (OSASS) within its Office of Research and Development. The *Environmental Outlook* reports prepared by the OSASS provide information on environmental trends and emergent problems and thus in turn enhance the EPA's long-range-planning process.[26] The EPA also has started to employ such information in developing its annual, congressionally mandated, comprehensive five-year R&D plan for supporting the agency's regulatory mission.

The movement of regulatory agencies such as the EPA toward strategic planning creates an opportunity for natural exchanges of methodology between the public and private sectors. Indeed, the public sector may increasingly look to the private sector for planning techniques as it has for a variety of other managerial innovations. Corporate planning provides a useful, albeit limited, model for strategic planning in environmental management. Planning principally involves:

Establishment of a planning horizon (the most distant point that can be observed in present trends),

Enactment of procedures to clarify goals and objectives,

Conduct of basic research and market forecasting,

Development of a range of future scenarios,

Identification of alternative courses of action,

Design of criteria by which to evaluate alternatives,

Assessment of existing programs and products,

Creation of policy recommendations.[27]

These activities are iterative and thus repeated in an abbreviated form in the programming and implementation stages of environmental management. More important, however, as Brita Schwarz points out, is the fact that "no complete analogy" exists for corporate planning in government organizations. This is because the public sector assumes that goal clarification is the responsibility of political forces.[28] Thus, the process of adminisstrative goal clarification undoubtedly will remain uneven and dependent on public hearings, risk and sensitivity analyses, opinion surveys, and general social-impact and technology assessments.[29]

The Future of Environmental Protection

That the private sector will deliver various social and environmental amenities, or at least exhibit greater social responsibility, is a central assumption guiding the regulatory-reform proposals of the Reagan administration. Whether or not government can devise proper incentives and the private sector assumes social responsibility for environmental management, it seems clear that the future of environmental protection depends upon increased public- and private-sector cooperation.[30] Harbingers of such action may be in the offing. For example, the social reporting systems emerging in European corporations, as well as selected examples of greater environmental sensitivity in U.S. firms, suggest in a limited fashion that more corporate social responsibility may be exhibited in the foreseeable future.

Whether or not the United States is on the verge of a new era of corporate environmentalism, the previous approach to environmental management would seem to be on the wane. It remains to be seen what the private sector will do with its increased freedom from regulatory constraints. Similarly, one cannot yet conclude with any degree of certainty that reducing the more-complicated procedural aspects of the EPA's regulatory system will produce clear benefits in terms of improved environmental quality as opposed to economic costs of compliance. While procedural simplification ought to foster economic efficiency with regard to compliance, simplification may not substantially alter degrees of compliance with environmental regulations. It is relatively clear, however, that existing procedures have not been a source of comfort for either the advocates or critics of environmental-protection measures. While the concepts we discuss do not necessarily guarantee a better situation, they might at least lead to a set of processes for addressing the current dissatisfaction.

Furthermore, concepts particularly associated with strategic planning and technology assessment can have their impact. The EPA was created on the basis of concepts that were intended to avoid capture while allowing the agency to address complex scientific and technical issues. These same concepts have, unfortunately, unduly limited the EPA's ability to pursue innovative approaches to defining and carrying out its protection mission. In essence, the traditional assumptions underlying public administration have been inappropriate for the highly complex scientific task of environmental protection.

Given continued demands for environmental protection, as well as pressures for deregulation and the inherent conflicts involved between the two, it is time both to evaluate the EPA's performance over the last decade and to redefine the agency's approach to environmental protection in light of that experience. Strategic planning, relying on techniques such as technology assessment and forecasting, represents one approach to anticipating rather than merely attempting to cope with these emergent conditions. It also provides an avenue for ex ante identification of the trade-offs inherent in shifting environmental priorities within the broader context of national goals and objectives. Such introspection becomes particularly valuable as a means for increasing the likelihood of continued improvements in environmental quality because market forces will have a greater influence on patterns of industrial expenditures for pollution control while the federal government will not be able to sustain the relatively high level of technology R&D funding experienced throughout the 1970s. Moreover, environmental gains are going to be less dramatic in the years ahead since fewer large uncontrolled sources (emitters) exist now than ten years ago. Thus, planning becomes all the more important to insure that attainable goals do, in fact, become realized.

Notes

1. For an overview of those concepts, see Philip Selznick, *TVA and the Grass Roots* (Berkeley: University of California Press, 1949); Kenneth Culp Davis, *Administrative Law Treatise* (St. Paul, Minn.: West, 1958); and Theodore Lowi, *The End of Liberalism* (New York: W.W. Norton, 1969), see especially chapter 10.

2. Paul W. MacAvoy, *The Regulated Industries and the Economy* (New York: Norton, 1979), see especially chapter 3; Simon Ramo, "Regulation of Technological Activities: A New Approach," *Science* 213 (August 21, 1981):837–842; and Lester B. Lave and Gilbert S. Omenn, *Clearing the Air: Reforming the Clean Air Act* (Washington, Brookings Institution, 1982).

3. National Academy of Sciences, *Decision Making in Environmental Protection Agency* (Washington: National Academy of Sciences, 1977).

4. Alfred Marcus, "Environmental Protection Agency" in *The Politics of Regulation* ed. James Q. Wilson (New York: Basic Books, 1980).

5. Ibid, p. 280.

6. Quoted in U.S. General Accounting Office, *Federal–State Environmental Programs—The State Perspective.* CED–80–106 (Washington: General Accounting Office, 1980), p. 43.

7. Lave and Omenn, *Clearing the Air,* p. 1. Lave and Omenn are cor-

rect in their assertion that the stationary source provisions of the Clean Air Act focus primarily on emission standards for new sources rather than the performance of existing ones. Similarly substantial shifts to natural gas and lower sulfur-content residual oil occurred from 1960 to the late 1970s. They also identify the limitations of existing mobile sources (automobiles) inspection and maintenance programs. While those assertions are accurate, significant emission reductions—particularly, sulfur oxide and carbon monoxide—did occur between 1970 and 1980; it is likely the rate of improvement would have been lower without the Act.

8. Louis Harris, "Testimony before the House Committee on Energy and Commerce, Subcommittee on Health and Environment," November 1981.

9. See James L. Regens, "Energy Development, Environmental Protection and Public Policy," *American Behavioral Scientist* 22 (November/December 1978):175–190. A 1982 Opinion Research Corporation nationwide poll conducted for the U.S. Chamber of Commerce found that "81 percent of the public and 78 percent of environmental activists agree that changes in the Act (Clean Air Act) probably can be made so that air quality will be protected at a lower cost than now." See *National Coal Association News,* No. 4601 (December 14, 1981):1.

10. Executive Order 12291 issued by President Reagan on February 17, 1981, requires the use of benefit-cost analysis and subsequent regulatory impact review by the Office of Management and Budget (OMB) for all significant regulatory actions, including environmental protection, unless prohibited by statute. A three-part test is used to define proposals as major rules subject to OMB review: (1) impact exceeding $100 million; (2) major increase in costs or prices; or (3) significant adverse effects on competition, employment, investment, productivity, or innovation.

Economic feasibility becomes a particularly thorny subject when the benefits appear highly intangible while compliance costs seem concrete. See Edward J. Mishan, *Cost-Benefit Analysis* (New York: Praeger, 1976); Robert Halvorsen and Michael G. Ruby, *Benefit-Cost Analysis of Air-Pollution Control* (Lexington, Mass.: Lexington Books, D.C. Heath and Company, 1981); and James L. Regens, "Equity Issues and Wilderness Preservation: Policy Implications for the Energy-Environment Tangle," in *Environmental Policy Implementation* ed. Dean E. Mann (Lexington, Mass.: Lexington Books, D.C. Heath and Company, 1982).

11. Guy Benveniste, *Regulation and Planning: The Case of Environmental Politics* (San Francisco: Boyd and Fraser, 1981).

12. Ibid., p. 111.

13. Richard Brooks, "The Legalization of Planning," *Administrative Law Review* (Winter 1979):31.

14. See Larry E. Susskind, "The Use of Negotiation and Mediation in Environmental Impact Assessment." Paper presented at the 1980 Confer-

ence of the American Association for the Advancement of Science (San Francisco); and Michael Lesnick and James Crowfoot, *Bibliography for the Study of Natural Resource and Environmental Conflict Resolution* (Ann Arbor: University of Michigan, School of Natural Resources, 1981).

15. These arguments are probably best summarized in Charles C. Schultze, *The Public Use of Private Interest* (Washington: Brookings Institution, 1977). Additional examples from the resource economics literature include A. Myrick Freeman, R.H. Haveman, and Allan Kneese, *The Economics of Environmental Policy* (New York: John Wiley and Sons, 1973); Allen Kneese and Charles Schultze, *Pollution, Prices and Public Policy* (Washington: Brookings Institution, 1975); and Edwin S. Mills, *The Economics of Environmental Quality* (New York: W.W. Norton, 1978).

16. For an elaboration of this concept, see Barry M. Mitnick, "Incentive Systems in Environmental Regulation," *Policy Studies Journal* 9 (Winter 1980):379–392; also note his, *The Political Economy of Regulation: Creating, Designing and Removing Regulatory Forms* (New York: Columbia University Press, 1980).

17. See Frederick R. Anderson, *Environmental Improvements Through Economic Incentives* (Baltimore: Johns Hopkins University Press, Resources for the Future, 1977), see especially pp. 145–191.

18. Garrett Hardin, "The Tragedy of the Commons," *Science* 162 (December 13, 1968):1243–1248.

19. Benveniste, *Regulation and Planning,* pp. 3–34.

20. David Wilson, *The National Planning Idea in the U.S. Public Policy: Five Alternative Approaches* (Boulder, Colo.: Westview, 1980).

21. The systems approach involves the use of basic biological metaphors to explain interactions between an organization and its environment. Systematic planning is designed primarily to improve the flow of survival signals, goal-realization feedback, and so forth and/or to limit the amount of entropy in the system. Major contributions to this planning approach include James Hughes and Lawrence Mann, "Systems and Planning Theory," *AIP Journal* 35 (1969):330–333; Anthony Catanese and Alan Steiss, *Systematic Planning: Theory and Application* (Lexington, Mass.: Lexington Books, D.C. Heath and Company, 1970); Alice Rivlin, *Systematic Thinking for Social Action* (Washington, D.C.: The Brookings Institution, 1971); Russell Ackoff, *Redesigning the Future: A Systems Approach to Society* (New York: John Wiley, 1974); Erich Jantsch, *Design for Evolution* (New York: Braziller, 1975); Frederick Emery and Eric Trist, *Towards a Social Ecology* (New York: Plenum Press, 1975); Gregory A. Daneke and Alan Steiss, "Planning and Policy Analysis for Public Administrators," in *Management Handbook for Public Administrators,* John W. Sutherland, ed. (New York: Van Nostrand Reinhold, 1978); and Kenneth Boulding, *Ecodynamics* (Beverly Hills: Sage Publications, 1979).

22. The adaptive-learning approach is concerned with planning as a

process of future responsive social learning and aims at procedures that facilitate the collective creation of alternative futures. Some of the major contributions to this approach include Donald Schon, "Maintaining an Adaptive National Government," in *The Future of U.S. Government: Toward the Year 2000* Harvey S. Perloff, ed. (Englewood Cliffs, N.J. Prentice-Hall, 1971); Donald Schon, *Beyond the Stable State* (New York: W.W. Norton, 1973); Donald Michael, *On Learning to Plan and Planning to Learn* (San Francisco: Jossey-Bass, 1973); and John Friedman, *Retracking America: A Theory of Societal Planning* (New York: Doubleday, 1973).

23. See Rensis Likert, *New Patterns of Management* (New York: McGraw-Hill, 1961); Paul Lawrence and Jay Lorsh, *Organization and Environment* (Cambridge: Harvard Business School, 1967); Warren Bennis and Philip Slater, *The Temporary Society* (New York: Harper & Row, 1968); Bertram Gross, "Planning in an Era of Social Revolution," *Public Administration Review* 31 (1970):259–296; Charles Perrow, *Organizational Analysis: A Sociological View* (Belmont, Calif.: Wadsworth, 1970); Chris Argyris, *Management and Organizational Development* (New York: McGraw-Hill, 1970); Eric Trist, "Developing Adaptive Planning Capability in Public Enterprise," in *Management Handbook for Public Administrators,* John W. Sutherland, ed. (New York: Van Nostrand Reinhold, 1978); Alan Steiss and Gregory A. Daneke, *Performance Adminstration* (Lexington, Mass.: Lexington Books, D.C. Heath and Company, 1980); and Robert Simmons, *Achieving Human Organization* (Malibu, Calif.: Daniel Spencer, 1981).

24. Michael, *On Learning to Plan,* pp. 255–280.

25. The two major technology assessments sponsored by the EPA were the Ohio River Basin Energy Study and the Technology Assessment of Western Energy Resource Development. Irvin L. White, *Energy from the West: Policy Analysis Report,* Environmental Protection Agency, Office of Energy, Minerals, and Industry, EPA–600/7–79–083 (Washington, D.C.: Government Printing Office, 1976). The Office of Energy, Minerals, and Industry has been replaced by the Office of Environmental Engineering and Technology within the Office of Research and Development.

26. For examples of initial attempts to provide such a capacity, see Environmental Protection Agency, Office of Research and Development, *Environmental Outlook 1980,* EPA–600/8–80–003 (Washington, D.C., 1980); and James L. Regens and David A. Bennett, "Environmental Quality Effects of Alternative Energy Futures: A Caveat on the Use of Macro-Modeling," in *Beyond the Energy Crisis,* Rocco Fazzolare and Craig Smith, eds. (Oxford: Pergamon, 1981).

27. Similar lists can be found in the basic textbooks of corporate strategic planning. Note, for example, William King and David Cleland, *Strategic Planning and Policy* (New York: Van Nostrand Reinhold, 1978); Peter Lorange and Richard F. Vancil, *Strategic Planning Systems* (Englewood

Cliffs, N.J.: Prentice-Hall, 1977); and Hans Thorelli, *Strategy + Structure = Performance* (Bloomington: Indiana University Press, 1977). For a review of public-sector efforts, see Michael Maskow, *Strategic Planning in Business and Government* (New York: Committee on Economic Development, 1978).

28. Brita Schwarz, "Long Range Planning in the Public Sector" *Futures* 9 (April 1977):116.

29. For a discussion of these tools and approaches and how they might be better integrated, see *Proceedings of the Conference on Public Involvement and Social Impact Assessment* ed. Gregory A. Daneke (Tucson, University of Arizona, College of Business and Public Administration, 1981).

30. For a further discussion, see Gregory A. Daneke and John Ehrmann, "Environmental Management in an Era of Deregulation and Reindustrialization" (Paper presented at the Conference of the American Society for Public Administration, Detroit, 1981).

The Courts, Technology Assessment, and Science-Policy Disputes

David M. O'Brien

The Judiciary as a Forum for Public Policymaking

Courts have historically, and especially since the late nineteenth century, adjudicated disputes arising from industrialization and technological developments. Indeed, as the senior circuit judge for the District of Columbia Court of Appeals observed, "Long before regulatory agencies, our system of private lawsuits served to control risk-taking. Consciously or unconsciously, society used that system to encourage some activities and discourage others."[1] After more the three decades on the federal bench, Judge David Bazelon surmised that "courts, in deciding private lawsuits, have responded creatively to harms that modern science has either produced or uncovered."[2] Whether courts have indeed *creatively* responded to science-policy disputes, whether and to what extent they may legitimately do so, and whether courts are *institutionally equipped and capable* of auspiciously responding remains problematic. Nonetheless, the judiciary has increasingly been drawn into complex science-policy disputes. Judges have assumed an important political role in reviewing and structuring assessments of the risks, costs, benefits, and responsibilities of developing technologies. In recent years, courts have confronted litigation involving computer technology and claims of personal privacy;[3] the patentability of organic life forms as a result of biomedical technology;[4] the occupational health and safety of workers exposed to allegedly hazardous substances such as cotton dust;[5] charges of environmental pollution;[6] and the scientific basis for regulations governing, and private compensation for, exposure to toxic and carcinogenic chemicals such as dichlorodiphenyltrichloroethane (DDT),[7] polychlorinated biphenyls (PCBs),[8] vinyl chloride,[9] benzene,[10] diethylstilbestrol (DES),[11] asbestos,[12] and radioactive wastes from nuclear reactors.[13] The impetus for judicial intervention in these disputes appears to be threefold: (1) the litigiousness of the American people, (2) congressional and administrative activities during the 1970s, and (3) judicial self-perception cum responsiveness to social forces.

Courts are not self-starters—they must passively await lawsuits raising actual cases or controversies—and therefore the litigiousness of the American people is significant, as commentators have noted since Alexis de

Tocqueville.[14] In 1980, lawsuits were being filed at the rate of 5 million a year in the fifty-state judicial systems and the local courts of the District of Columbia and Puerto Rico, while no less than 188,487 suits were filed in federal district courts, and the work load of federal courts of appeals rose 39.3 percent over that of 1975.[15] "It seems that the less [the American people] participating in voting," J. Woodford Howard comments, "the more they litigate in the belief, however mistaken, that law is a fit instrument of social change and that federal courts offer cheaper, faster, and more effective solutions to their problems."[16]

The judicial forum appears to be authoritative and responsive. Litigation, for individuals frustrated or thwarted by the political process, serves not merely as a way of achieving private compensation but as a vehicle for direct participation in the formulation of public policy. Private-law adjudications, Milton Katz points out, "offer a means whereby the multiple initiatives of private citizens, individually or in groups, can be brought to bear on technology assessment, the internalization of costs, and environmental pollution."[17] Courts are emerging as forums for technology assessment partially because of the pressures of litigation brought by individuals and special-interest groups seeking either compensation for personal injuries or the prevention of ostensibly detrimental health, safety, and environmental consequences of developing chemical and industrial processes.

Judicial intervention has also been invited by the expansion of congressional legislation and administrative regulations on health, safety, and environmental matters. There are, for example, some 31 federal statutes, passed primarily during the 1970s, governing toxic and carcinogenic substances.[18] Health, safety, and environmental regulations precipitate litigation. The number of civil suits challenging regulations and enforcement proceedings, encountered by the EPA, for instance, rose from less than 20 in 1973 to almost 500 in 1978.[19] Special-interest groups are afforded numerous opportunities for challenging the scientific basis for agency regulation—for example, the Occupational Safety and Health Administration's (OSHA) regulation of cotton dust[20]—as well as agency failure to regulate—for example, Consumer Protection Safety Commission's ban on TRIS, an allegedly carcinogenic fire retardant sprayed on baby clothes.[21]

As a consequence of congressional and administrative activities during the 1970s, the judiciary, on the one hand, no longer occupies the central place that it once ostensibly did in resolving science-policy disputes through private-law adjudication. Federal courts, on the other hand, more frequently confront complex science-policy issues as a result of their expanded role in supervising federal-agency rule making and adjudication. The allure of a judicial/administrative partnership in health, safety, and environmental regulation forged a "new era in administrative law,"[22] and it promoted litigation over issues on "the frontiers of science" and technology.[23]

The judiciary has itself encouraged both private-law adjudication and challenges to administrative regulations involving scientific and technological disputes. Courts have made their own assessments of the risks and responsibilities of developing technologies and thereupon extended traditional common-law doctrines as well as increasingly reviewed and challenged the process and scientific basis adduced for administrative regulations. The institutional limitations of courts and judges' lack of scientific expertise have undoubtedly discouraged to some degree judicial intervention in science-policy disputes, even while legal realism has contributed to judicial activism on civil-liberties/civil-rights matters and intervention in the operation of public schools, prisons, and mental-health institutions.[24] Still, given the pressures of private-law litigation and the mounting challenges to federal regulation, state and federal courts have, even if reluctantly, responded and in turn contributed to the perception of the judiciary as a forum for public-policy formulation.

Ironically, the more courts are responsive, the more they potentially threaten their legitimacy as dispassionate, authoritative, and conclusive tribunals, especially when extensive media attention is devoted to science-policy disputes. The plight of federal district-court Judge Miles Lord in the controversy over the Reserve Mining Company's daily dumping of 67,000 tons of taconite (a low-grade iron ore) into Lake Superior is extreme yet illustrative. In litigation lasting a decade, Judge Lord became immersed in the technical details of taconite mining and the problems of assessing the carcinogenic risk of taconite. He also became outraged by Reserve Mining's diversionary legal tactics and the reluctance of local political officials to intervene because of their fear of economic loss to the community if the company closed. During the course of the trial, Judge Lord, likening himself to Moses in the wilderness, expressed his frustration:

> Now it's very difficult to be a[]lone and lonesome federal judge at a time like this, because it affects the lives of many people. Many people who are working versus many people who may be dying. . . .

> I personally question the wisdom, and I make this observation to you, of even having a Minnesota judge sit on this [case], and particularly question the wisdom of having a local judge . . . sit on such a case, because the pressures are just too much.[25]

Impatient with the accusatorial process for settling science-policy disputes, Judge Lord called and examined expert witnesses and offered his own testimony as well as stated that defense witnesses could not be trusted, that "in every instance and under every circumstance, Reserve Mining Company hid the evidence, misrepresented, delayed and frustrated the ultimate conclusions."[26] The Court of Appeals for the Eighth Circuit eventually removed him from the litigation, remarking, "Judge Lord seems to have shed the robe of a judge and to have assumed the mantle of the advocate. The court thus becomes a lawyer, witness and judge in the same proceeding and abandons the great virtue of a fair and conscientious judge—impartiality.''[27]

The Reserve Mining controversy illuminates some of the inherent tensions in the judiciary as a forum for technology assessment and for resolution of science-policy disputes. Science-policy disputes are frequently polycentric, or many sided.[28] They are characterized by considerable scientific uncertainty and intense value conflict that transcends the parties joined in the litigation and, hence, entail complex socio-economic and political repercussions for a community. Polycentric disputes are not auspiciously resolved in the judicial arena. When courts are confronted with polycentric disputes, as in the Reserve Mining case, adjudication tends to be lengthy, costly, and inconclusive. No less significant is the fact that courts of appeals are removed from the local political pressures surrounding trials, whereas federal and state trial judges must come to terms with the political pressures and ramifications of their decisions. Polycentric disputes seemingly confront trial courts with two unattractive alternatives. Some judges, like Miles Lord, may feel pushed to innovate procedurally, directly participating in the dispute and stating an unambiguous preference for the outcome. More typically, as trial attorneys often complain about state judges in health and environmental litigation, they tend to formulate issues and admit only evidence conducive to the accusatorial process but less amenable to resolution of the dispute.

Private-Law Adjudication: The Regulatory Function of Courts

Private-law adjudication is an important, often underestimated, social force. The public nature of private-law litigation arises from providing remedies ''for the everyday hurts inflicted by the multitudinous activities of our society.''[29] Tort law, as Leon Green perceptively remarked, is ''public law in disguise.''[30] Private law serves as a catalyst for the marketplace and as a vehicle for public policymaking. In defining rights and remedies and in balancing equities in cases involving, for example, environmental pollution, courts inevitably assume a regulatory role.

There are, prima facie, distinct advantages for individuals and special-interest groups litigating disputes over industrial air pollution, disposal of hazardous substances, and the introduction of carcinogens into the environment. Although the economic and informational costs of adjudication often exceed the resources of would-be litigants,[31] they are less than the comparable economic and social-transaction costs of organizing and mobilizing political forces in legislative and regulatory arenas, even though for a would-be litigant the costs of collective action would be lower.[32] Litigation is also advantageous in focusing on particular facts and claims in a more-discriminating fashion than legislation that, for instance, imposes a tax on effluents. Litigation holds the potential for direct compensation, whereas regulation need not and a tax scheme may not address entire categories of injuries.[33]

The role of private-law litigation in shaping market decisions and assessing technological developments only received attention since Arthur Pigou introduced the idea of market transactions having external social costs.[34] Competitive pressures for investments and innovations in the marketplace discourage businesses from taking into account the externalities of their activities—the secondary effects of, and the impact on, parties not directly involved in economic transactions. Externalities like air pollution may accompany the production of a good but not be reflected in its price as determined by market forces. A chemical plant that ignores pollution will have lower costs than one that treats its effluents and thus will prosper at the expense of its more environmentally responsible competitors. No plant could afford indefinitely to assume (internalize) the costs of pollution control that its competitors did not likewise assume. While pollution control might be achieved by government regulation, subsidy schemes, or taxes on effluents, the assertion of legally recognized rights also forces plants to assume the costs of pollution control. The prospect of liability for pollution creates incentives for the plant and the industry to conduct activities in a safer—or, at least, legally defensible—manner. Private-law litigation, as public-interest groups such as the Environmental Defense Fund and the Natural Resources Defense Council appreciate in bringing so-called tests cases, thus becomes a vehicle for structuring technological decision making and for securing public goods such as clean air.[35]

In 1980, the Supreme Court let stand a decision by the California State Supreme Court that dramatically illustrates the potential impact of private-law adjudication. In *Sindell* v. *Abbott Laboratories,* Mrs. Sindell and other women with vaginal disorders filed suits against manufacturers of DES, alleging that their disorders resulted from the use of DES by their mothers.[36] DES, a synthetic estrogen, has been manufactured since 1941 by approximately 200 companies, with FDA approval, as a prescription drug for preventing miscarriages. Under traditional tort doctrine, plaintiffs must demonstrate negligence on the part of the defendant, their personal injury, and a causal relationship between the defendant's action and the injuries suffered. Mrs. Sindell and the other women were unable to meet the latter criteria since it was impossible to conclusively identify which company had manufactured the DES taken by the women's respective mothers. The California State Supreme Court, however, set forth a new theory for recovery. A plaintiff who cannot identify the tort-feasor, the court held, may sue the manufacturers who collectively represent a substantial share of the market. Moreover, if the plaintiff shows that the defendant acted negligently and that the product caused injury, then the burden shifts to each defendant to prove that its specific product could not have caused the injury. Liability, like causation, the court concluded, should be assigned on the basis of each defendant's market share. *Sindell*'s market-share theory and imposition of joint liability significantly expands the liability of the pharmaceutical industry for injurious products. For manufacturers of DES, claims for recovery could total $40 billion.[37] For the pharmaceutical industry, the decision bodes an increase in litigation costs as companies may have to defend themselves in a larger number of lawsuits, even if they have only a small share of the market. *Sindell* provides an incentive for the industry to conduct more-searching assessments of the effects of new drugs, and therefore it may affect future technological innovations, delay or preclude introduction of some new drugs, and increase the cost of liability insurance and drug prices.

By creating and defining the contours of rights and remedies, courts influence the marketplace and assessments of the potential consequences of developing technologies. As Laurence Tribe observes:

> Whether one is dealing with a technological process as simple as that of cattle-grazing or one as complex as that of computerized information-processing, the law can create a market by "creating rights"—in the former case, be defining the "property" rights of the owner of a field so as to incorporate a power to exclude uninvited cattlegrazers; in the latter case, by defining the "personal" rights of the subject of an information system so as to incorporate a power to exclude uninvited users and unauthorized uses of "private" data.[38]

When enforcing personal and proprietary rights at common law, courts historically have influenced the marketplace. Judges also explicitly or implicitly make their own assessments of the impact of technology by balancing the severity of damages with the utility of particular technological and industrial processes.

In historical perspective, the constraints of legal reasoning and doctrines, especially the emphasis on property rights, foreclosed the possibility of recovery for health and environmental damages. Property owners always had a right to compensation for damages by contiguous property owners. Yet until the mid-twentieth century, courts refused compensation for noise and invisible-gas pollution because traditional trespass doctrine embraced only the uninvited presence of corporeal objects on an individual's property.[39] Throughout the nineteenth and twentieth centuries, actions for health, safety, and environmental injuries were no less constrained by doctrines such as the contributory negligence of the plaintiff, the assumption of risk in a situation—for example, a worker's taking a dangerous job—and the proximate cause between a plaintiff's injury and the defendant's allegedly harmful action.

Nevertheless, in the last thirty years, a significant evolution in private-law adjudication has occurred, a movement Jethro Lieberman claims toward fiduciary duties: "The response of the legal system to the movement toward a welfare state has been to repudiate the classical liberal doctrine *caveat emptor* (let the buyer beware) and related rules. In their place, the law is imposing ever more stringent duties of care on those who act."[40] Courts have indeed permitted greater access to the judicial arena and modified legal doctrines, thereby awarding in some cases compensation for health, safety, and environmental damages.

The Courts and Toxic Torts

When industrial activities interfere with individuals' personal or proprietary interests, they may sue under several legal doctrines such as trespass, nuisance, and strict liability for abnormally dangerous activities. Actions for trespass, unlike those for nuisance, involve liability for interference with an individual's right exclusively to own and enjoy property. Nuisance, however, proves a more-flexible doctrine in environmental litigation.[41] Although the law of each state governs actions for nuisance, essentially plaintiffs must show an actual injury—that reasonable enjoyment of property was interfered with or that health was endangered—and that the defendant's activities were the proximate cause of the injury.[42] Individuals may sue for private nuisance when substantial and material interference with their property results from, for example, industrial dust, noise, odors, and water pollution. Alternatively, they may bring actions for public nuisance when an activity involves an unreasonable interference with a right common to the general public and when the alleged injuries are substantial and widespread. The Reserve Mining controversy involved claims that dumping taconite created a public nuisance, and courts have similarly found activities such as acid mine drainage,[43] blasting operations in limestone quarries,[44] and disposal of hazardous wastes to constitute public nuisances.[45]

The Reserve Mining controversy was also one of the first cases raising claims to a federal common law of nuisance, which was not recognized until the 1970s. In 1972 the Supreme Court, in *Illinois* v. *City of Milwaukee,* indicated that lower federal courts had the power to create federal common law concerning the pollution of interstate and navigable waters when there was an overriding interest in a uniform rule or when a controversy touched on basic interests of federalism.[46] Industrial water pollution thus could ostensibly be challenged for creating a public nuisance under state and/or federal common law. Federal common-law litigation, however, was soon overshadowed by the enactment, just six months after the Court's decision, of the Federal Water Pollution Control Act (FWPCA) amendments of 1972.[47] The FWPCA permits citizen suits against industrial polluters, and federal agencies now concentrate on implementation of the statute rather than costly and laborious actions against individual dischargers. Some lower courts nevertheless maintained not only that the FWPCA did not preempt federal common law but also that the federal common law of nuisance could be asserted by local governments,[48] and they allowed judicial awards of compensatory damages even where industries had abated discharges,[49] as well as suggested extension to actions for air pollution and disposal of ultrahazardous wastes such as in the Love Canal controversy.[50] The political consequences of an open-ended federal common law of nuisance and the potential for a "government by the judiciary" apparently impressed a majority of the Supreme Court.[51] Nine years after *Illinois* v. *City of Milwaukee,* The Court qualified its decision by holding that lower federal courts may create substantive common law only in narrow, restricted instances and that when federal statutory law like the FWPCA governs a question previously the subject of federal common law, the courts must assume that the congressional legislation is controlling. Justice William Rehnquist, writing for the majority in *City of Milwaukee* v. *Illinois,* underscored that the federal, unlike the state, judiciary possesses only limited jurisdiction and has no general powers to develop and apply its own common-law rules.[52] Federal judges, moreover, have neither expertise nor political legitimacy in fashioning general rules for resolving complex disputes such as those involving environmental pollution. He further noted the problematic nature of allowing localities and states to challenge out-of-state discharges in federal courts and to impose federal common-law standards that are more stringent than comparable congressionally enacted standards.[53] By contrast, Justice Harry Blackmun, for the three dissenters, argued that federal common law complements congressional action in furthering interests embodied in the Constitution or federal legislation, and he observed that "each State [has] the right to be free from unreasonable interference with its natural environment and resources when the interference stems from another state or its citizens."[54]

Potentially more important for environmental-tort-litigation are the changes forged in the doctrine of strict liability, or liability without fault for abnormally dangerous activities. The doctrine of strict liability derives from an 1865 English common-law case, *Rylands* v. *Fletcher.*[55] The case involved damages due to a reservoir, erected on an abandoned coal mine, that leaked water through an unused shaft and eventually flooded Rylands's coal mine. The Exchequer Chamber established the principle of strict liability, stating:

[T]he true rule of law is that the person who for his own purposes brings on his lands and collects and keeps there anything likely to do mischief if it escapes, must keep it at his peril and if he does not do so, is prima facie answerable for all the damage which is the natural consequence of its escape.[56]

Subsequently, in the United States, the first and second *Restatement of Torts* adopted the principle of strict liability for "extrahazardous" and "abnormally dangerous" activities.[57] Accordingly, state courts have permitted compensation for damages resulting from smoke, dust, and noxious gases emitted from industries;[58] fumigation with cyanide gas;[59] crop dusting;[60] and exposure to toxic chemicals.[61] *United States* v. *FMC Corporation* is an intriguing illustration of the extent to which some courts have gone when applying the doctrine of strict liability.[62] FMC Corporation manufactured a highly toxic pesticide, carbofuran. Residues of the pesticide were deposited on the company's property in a pond that also served as a site for migratory birds. Reasoning from *Rylands*, District Court Judge John Curtin ruled, as was upheld by the court of appeals, that FMC Corporation was liabile for killing the birds since it failed to prevent the toxin from reaching the pond, even though it was unaware of the lethal quality of the water of the pond.

Health and Safety and Hazardous Products

Along with the incremental changes forged in doctrines of trespass, nuisance, and strict liability for abnormally dangerous activities, courts have modified theories of negligence and strict products liability when confronted with disputes over allegedly hazardous products. Both of these theories may impressively expand private-law litigation over toxic and carcinogenic substances.[63]

Negligence actions fundamentally center on the question of whether a company failed to exercise due care in manufacturing and marketing a product. More specifically, as in the Michigan episode over polybrominated biphenyl (PBB), individuals may assert that a company was negligent in failing to properly inspect a product and that the lack of due care was the proximate cause of injuries.[64] In the Michigan controversy, a company mistakenly shipped to a farm-feed supplier bags of a fire retardant, Firemaster, containing PBB, instead of a cattle-feed additive, Nutrimaster. The fire retardant was subsequently fed to cattle and resulted in widespread contamination of the cattle and eventually the general population of Michigan; more than 90 percent of Michigan residents have PBB in their bodies, a Mount Sinai research team reported in 1979.[65]

In other situations, individuals may alternatively argue that companies were negligent in failing either to warn of the potential hazards and known defects of their products or properly to design and test their products.[66] In *Boyl* v. *California Chemical Company,* for instance, Mrs. Evelyn Boyl won compensation for injuries suffered from exposure to sodium arsenite in a weed killer, Triox, she had purchased and used.[67] After spraying her patio with Triox, she dumped the residue and waste water from cleaning the spray tank on a portion of her backyard. A couple of days later she unwittingly took a sun bath on the spot where she had previously poured the Triox rinse. Shortly thereafter, she developed a heat rash, followed by dizziness, muscle tremors, and physical malfunctioning, for which she was hospitalized. Although California Chemical Company printed a warning, a poison skull and crossbones, and instructions for the use of Triox on every bottle, the federal district court ruled that the company was negligent in failing to provide a reasonable warning of the health risks of the solution and that such negligence was the proximate cause of the injuries. Mrs. Boyl, the court furthermore held, was not contributorily negligent inasmuch as she "was totally ingnorant of any risk or danger to herself arising from the Triox solution contaminated earth."[68]

Boyl indicates the potentially broad compass of the law of negligence and the role of courts in balancing the risks and benefits of consumer goods. Implicit in the litigation is the rather amorphous concept of unreasonable harm. Unreasonable harm is defined in *The (Second) Restatement of the Law of Torts* as any "risk that is of such magnitude as to outweigh what the law regards as the utility of the act or of the particular manner in which it was done."[69] Yet, when determining that a product poses an unreasonable risk of harm, judges and juries inescapably render their own assessments of the risks and responsibilities involved in each case. *Boyl, Sindell,* and other judicial evaluations of risk reveal some of the dynamics and political import of the judiciary as a forum for technology assessment. Judges, who are not directly politically accountable for their decisions, need neither articulate nor quantify the relative weight of the values underlying their balancing, even though their rulings and awards may vary greatly from one jurisdiction to another and may entail complex socioeconomic repercussions.

The potentially broad socioeconomic impact of judicial assessments of the reasonableness of health and environmental risks also heightened controversy over the extension of the doctrine of strict product liability.[70] In recent litigation, the doctrine embraced claims by workers for compensatory damages due to exposure to asbestos,[71] and it prompted the unsuccessful attempt by former members of the military in Vietnam to win recompense for injuries allegedly resulting from exposure to various phenoxy herbicides, including 2,4,5-T, so-called Agent Orange.[72] Although considerable debate exists over the differences between assessing product design under the negligence calculus of risk and more-modern product-liability standards, there is no gainsaying that judges have enlarged the doctrine of strict product liability in response to newfound risks associated with technological and industrial advances. The Model Uniform Product Liability Act (MUPLA) provides some guidance for judicial risk assessments: "In order to determine that the product was unreasonably unsafe in construction, the trier of fact must find that when the product left the control of the manufacturer, the product deviated in some material way from the manufacturer's design specifications or performance standards, or from otherwise identical units of the same product line."[73] In evaluating the social risks, costs, and benefits of a product, the act also suggests that consideration should be given to evidence such as the following:

(a) Any warnings and instructions provided with the product;

(b) The technological and practical feasibility of a product designated and manufactured so as to have prevented claimant's harm while substantially serving the likely user's expected needs;

(c) The effect of any proposed alternative design on the usefulness of the product;

(d) The comparative costs of producing, distributing, selling, using, and maintaining the product as designed and as alternatively designed;

(e) The new or additional harms that might have resulted if the product had been so alternatively designed.[74]

Judges are nonetheless relatively free to make their own assessments of technology and the feasibility of alternative technologies when imposing strict product liability. Strict product liability, even more than other tort theories, illuminates the regulatory role of judicial rulings. Judicial imposition of strict product liability forces industries to internalize the cost of injuries to consumers of a product by transferring the cost back to the industry. "The purpose of [strict] liability," California State Supreme Court Justice Roger Traynor notes, "is to insure that the costs of injuries resulting from defective products are borne by the manufacturers that put such products on the market rather than by the injured persons who are powerless to protect themselves."[75] The Eighth Circuit Court of Appeals even more candidly proclaimed that "the risk of personal injury has become a cost of doing business."[76] Indeed, courts frequently justify the doctrine of strict product liability "on the realization that our technological society, with its proliferation of products and mass advertising, demands judicial protection of the consumer who has neither the capacity nor opportunity to discover latent dangers in products.[77]

The Impact and Limitations of Private-Law Adjudication

Courts encourage private-law adjudication of science-policy disputes by extending common-law doctrines and by occasionally making substantial awards. In 1980, an estimated 10,000 suits were pending over damages due to exposure to asbestos alone, and residents of Love Canal in New York had begun approximately 1,300 actions seeking $15 billion in compensatory and punitive damages for exposure to hazardous-chemical wastes.[78] Media fascination with large awards like that made in the Karen Ann Silkwood case further encourages would-be litigants. Karen Silkwood, deceased, initially received $500,000 compensation and $10,000,000 punitive damages for exposure to escaped plutonium.[79]

Landmark cases and large awards are more suggestive than revealing about the trends and impact of private-law adjudication. Perhaps more significant is the fact that between 1977 and 1979 a third of the state legislatures reversed state-court rulings that made it easier for individuals to gain access to the judicial arena and to prove their claims of health and environmental damages.[80] Despite the potentially enormous economic impact of strict product-liability rulings, studies estimate the social costs of product-related injuries to be more than $10 billion a year, while fewer than one percent of those injured file a lawsuit.[81] Industries still win an estimated three-quarters of the cases that go to juries, and the average award is less than $4,000.[82] Selective, unsystematic imposition of tort liability also appears "to maldistribute benefits by overcompensating the slightly injured person and undercompensating the seriously injured."[83] Studies of disputes over toxic and hazardous substances likewise indicate immense disparity between the social costs of injuries and the number of lawsuits filed and the amount eventually recovered.[84]

In a number of significant tort actions, judges have assessed the social costs and benefits of technological advances and thereupon awarded compensation, forcing the private sector to internalize some of the costs of hazardous industrial processes and consumer products. Private-law litigation, however, remains an inauspicious vehicle for public-policy formulation and technology assessment. This is partially due to institutional limitations of the judicial forum. The accusatorial process as well appears neither appropriate for nor amenable to resolution of the scientific uncertainties underlying health and environmental litigation. The institutional limitations of courts and the procedural obstacles in the accusatorial system, in turn, result in distributive injustice in compensatory awards and fragmented, redundant, and often conflicting judicial rulings.

The institutional structure of courts appears to be ill-suited for resolution of complex science-policy disputes and for undertaking technology assessments, especially assessments of prospective damages. Courts are passive, reactive institutions. The judiciary possesses neither the informational nor the financial resources of Congress or the executive branch. Litigation focuses retrospectively, not prospectively, on a dispute; adjudication centers on actual injuries suffered; and legal issues are framed and information generated by adversaries.[85] In some instances judges have acknowledged that "exposure [to a carcinogen] is more in the nature of a continuing tort" but that prospective, secondary damages are not typically considered or relevant.[86]

Judges possess a potentially powerful instrument—the power to issue injunctions—by which they may address prospective health and environmental damages raised in litigation. Injunctions may be used to oblige a polluter to terminate a particularly hazardous technology or to force the polluter to find alternative technologies, as well as to encourage litigants to negotiate an agreement for compensation of damages caused by the employment of a particular technology. Nonetheless, judges are usually reluctant to intervene extensively in a science-policy dispute because of their lack of expertise in scientific and technical matters and due to the incremental nature of legal reasoning and judicial reliance on stare decisis, or precedent.[87] The institutional limitations of the judicial forum indicate not merely, as Harold Green concluded, "that we cannot rely on the courts alone to protect society against fast-moving technological developments."[88] Courts appear to be neither structurally nor situationally predisposed to resolving but a modest number of disputes over the impact of past technological developments, and judges lack the resources, opportunity, and training to assess the impact of advancing technologies.

Numerous practical and procedural problems arise with the adversary process as a mechanism for resolving complex science-policy disputes. In health and environmental litigation, the practical problems stem from the state of scientific knowledge and the art of medical diagnosis that not always permit conclusions about the causal link between a plaintiff's injuries and the defendant's activities. The cost of bioassays and the rigor of scientific proof often prevent generation of information adequate to support legal recoveries. Diseases often have multiple causes and long latencies, and therefore it becomes extremely difficult, if not impossible, to identify the exact chemical agent causing a disease or the particular polluter responsible. Most lawsuits are far more complex than a single individual's alleging that an industrial plant discharged a pollutant that caused a particular disease. More typically, many people in a community have been exposed, the disease has a latency period of 20–30 years, and a number of environmental variables intervene. While an increase in the incidence of a disease may be revealed statistically, no direct evidence of the disease may exist in humans even though strong or overwhelming evidence can be found in laboratory animals. Even when scientists agree on the interpretation of animal bioassays, they may profoundly disagree on the correct inferences to be drawn from the data. Scientific disagreement thus remains over whether a threshold exists for carcinogenic risk and how to extrapolate the carcinogenic effects of high-dose levels to low-dose levels.[89] The problems of proving causation become even greater in environmental litigation when there are several sources of pollution. Each polluter may individually cause the disease, but the possibility exists that only one does, even though there may be uncertainty as to which polluter. Alternatively, one polluter may not itself cause a disease yet, in combination with other polluters, may precipitate the disease. Finally, there is also the problem that individuals frequently live part of their lives by one industrial polluter that by itself does not pose a health hazard but then move to an area polluted by a different source, and the combined exposure leads to the disease.[90]

Scientific uncertainty over the risk of toxins and carcinogens is impossible to resolve within adversarial proceedings.[91] Rules of evidence limit the introduction of some forms of scientific data and testimony.[92] Statutes of limitations frequently foreclose even the possiblity of claiming injuries involving diseases with long latency periods, such as cancer.[93] Common-law defenses further compound the problems of litigating disputes involving scientific uncertainty. For instance, employers normally have immunity from negligence suits by employees who were exposed to chemical hazards and from liability for injuries resulting from risks that were unknown or are not yet factually supported by scientific data.[94] Private-law litigation appears to be particularly ill suited for resolution of science-policy disputes such as those over air and water pollution because the pollution effects are weakly associated with existing individuals and the damages are thinly spread among those individuals. "From the standpoint of society as a whole," Julian Juergensmeyer observes, "the most that can be expected from air pollution control through assertion of private rights is the handling of some instances of air pollution which cannot be or are not yet controlled by public regulation."[95]

Private-law adjudication of science-policy disputes is costly, cumbersome, and time-consuming. Adjudication in turn discourages lawsuits and promotes out-of-court settlements since litigation becomes economically unreasonable for individuals suffering and in need of reparations.[96] Consequently, private-law adjudication maldistributes compensation. Only a few injured individuals actually carry through with a lawsuit and even fewer win recovery for damages. The inequities in compensation also arise from the fact that the private-law adjudication, considered as a political process, fragments decision making. Within the fifty states and between state and federal jurisdictions, private-law litigation inevitably produces fragmented, piecemeal, and often redundant and conflicting decisions.

The Allure of a Judicial/Administrative Partnership

Political pressures, arising partially from concerns that neither the private sector nor private-law litigation effectively and equitably governs the social costs of technological and industrial processes, prompted Congress in the 1970s to enact major health, safety, and environmental legislation. Congress delegated broad powers to federal agencies to assess health, safety, and environmental risks and thereupon to promulgate extensive regulations. In some instances, as with the Clean Air Act, Congress also mandated technology-forcing requirements—that is, agencies are required to set, within certain limits, standards for pollution control commensurate with available or feasible technology and thus to "force regulated sources [that is, industries] to develop pollution control devices that might at the time appear to be economically or technologically infeasible."[97]

Political conflict envelops health, safety, and environmental regulations due to their complex socioeconomic ramifications and because they rest on scientific data that remain subject to considerable disagreement within the scientific community. The polycentric nature of and the scientific uncertainties underlying health, safety, and environmental regulations inexorably invite litigation. Moreover, unlike New Deal regulatory agencies, contemporary agencies such as the EPA are isolated neither from central political control nor extensive judicial review; administrative discretion and expertise are subject to heightened political suspicion and judicial scrutiny.[98] Whereas prior to World War II courts manifested considerable deference to federal agencies, contemporary federal courts intensively and extensively review both the process generating, and the substance of, federal regulations.[99]

Thus a new era in administrative law and regulation forged a partnership between the lower federal courts and the administrative agencies. While federal judges proclaim their role to be "that of a constructive cooperation with the agency involved in the furtherance of the public interest," the judicial/administrative partnership remains inevitably tenuous, often more acrimonious than amicable.[100] The federal judiciary nonetheless has an important participatory role in the formulation and implementation of federal regulation. Litigation has become part of the regulatory process. Agencies inescapably must defend all major regulations in judicial forums; federal regulation appears to require the judicial imprimatur.

Standing and Challenging Regulatory Action and Inaction

The decisions of federal agencies are usually subject to judicial review unless specifically precluded or committed to agency discretion by statute.[101] The

Administrative Procedure Act establishes a presumption of reviewability: "A person suffering legal wrong because of agency action, or adversely affected or aggrieved by agency action within the meaning of a relevant statute, is entitled to judicial review thereof."[102] To challenge administrative actions, individuals must show that they have standing to sue—that they have exhausted administrative appeals and that they have demonstrable personal interests recognized as within the "zone of interests" protected by a statute.[103] Health, safety, and environmental legislation typically provides both for a statutory right of review of regulatory action and for citizen suits as a means of enforcing legislative mandates.[104] When legislation does not provide for citizen suits, individuals may still claim a "private cause of action" under a statute to gain access to the courts and to force compliance with congressional objectives.[105]

Judicial intervention in regulatory politics has also been occasioned by the Supreme Court's liberalization of the law of standing, permitting greater access to the judicial arena and opportunities to assert a broad range of interests in challenging federal agencies. Historically, individuals acquired access to the courts only by demonstrating personal or proprietary damage and hence that they had standing to sue. The requirement of a personal or proprietary damage limited access to so-called pocketbook plaintiffs and precluded the use of private-law litigation to raise major issues of public affairs.[106] In the late 1960s, however, the Court held that judicial review of administrative decisions was a presumption of congressional legislation and expanded the permissible range of taxpayer suits. Subsequently, the Court indicated that individuals could challenge regulatory policies for infringing not only economic interests but also "aesthetic, conservational, and recreational" interests.[107] In *United States* v. *Students Challenging Regulatory Agency Procedures (SCRAP),* the Court underscored its approval for lower federal courts to entertain lawsuits based on claims to other than economic damages.[108] Here, a group of law students (SCRAP) challenged a proposed surcharge on railroad freight approved by the Interstate Commerce Commission. They argued that the eventual effect of the regulation would be damage to the environment since the surcharge would operate to discourage the transportation of recycled materials. In granting SCRAP standing, the Court observed:

> Aesthetic and environmental well-being, like economic well-being, are important ingredients of the quality of life in our society, and the fact that particular environmental interests are shared by the many rather than the few does not make them less deserving of legal protection through the judicial process.[109]

Although retaining the presumption that individuals demonstrate a personal injury, the Supreme Court remolded the law of standing so that plaintiffs may function as surrogates for special-interest groups. Individuals still must demonstrate that they have a personal stake in an actual case or controversy and a personal injury that is "distinct and palapable" and that a "fairly traceable" causal connection exists between the injury and the challenged regulation.[110] Nonetheless, the personal injury claimed actually embraces a public injury.

Thus the federal judiciary promises a forum for raising science-policy disputes and challenging regulatory action and inaction. Agencies encounter attacks for failing to comply with the NEPA,[111] for example, for adopting summary judgments,[112] or for not allowing adequate opportunities for interested groups to participate in agency rule making.[113] Even when interest groups are afforded opportunities to participate in rule-making proceedings, they may still challenge the final regulation on procedural and substantive grounds. Indeed, precisely because health, safety, and environmental regulation is often based on incomplete information, conflicting scientific data, and problematic interpretations thereof, and because it involves fundamental policy choices, agencies must defend their regulations.[114] Agencies typically must defend in litigation the scientific basis for their regulations because scientific data remain open to different calculations and interpretations. If agencies fail to negotiate agreement with the relevant interest groups on the scientific basis for their regulations, then they will certainly face a court battle. Litigation, albeit providing an opportunity for direct citizen participation in regulatory politics, proves costly, time consuming, and potentially threatens to undermine administrative legitimacy since federal regulation appears to indeed turn on the judicial imprimatur.

Substantive and Procedural Review: Science-Policy Disputes and the Judicial/Administrative Partnership

Rule making is central to regulation, and although only a small percentage of agency decisions is reviewed by the courts, judicial review significantly structures and checks administrative rule making. Judicial review does not in itself ensure the correctness or even the fairness of administrative rule making. Instead, judicial review basically serves to ensure that agencies do not exceed their delegated authority, deny individual's constitutionally or statutorily protected rights, or abuse their discretion by arbitrarily and capriciously conducting their proceedings or supporting their decisions. Unless specified in enabling legislation, federal agencies have flexibility in adopting either informal (notice and comment) procedures or formal, adjudicatory-type rule-making procedures under the Administrative Procedure Act. The Administrative Procedure Act also specifies that informal rule making should be set aside only when a court finds agency action to be "arbitrary, capricious, an abuse of discretion, or otherwise not in accordance with law"; whereas regulations based on formal rule-making procedures are subject to more-rigorous review, a demonstration of substantial evidence on record.[115]

Federal agencies usually prefer informal rule-making procedures.[116] Informal rule making appears less costly, less time consuming, and ostensibly more efficient and productive when regulation involves complex, polycentric science-policy disputes. Formal, trial-type procedures are no more beneficial for resolving science-policy disputes in the administrative arena than private-law litigation in the judicial arena. Adjudicatory processes cannot accommodate the polycentric nature of the issues, the irreducible factual uncertainties cum scientific controversies, and the numerous and diverse interests in the dispute. Expansion of public participation and extension of adjudicatory-type procedures—for example, presentation and cross-examination of expert witnesses—frustrates rather than enhances administrative fact finding and decision making.

In the last decade, when confronted with litigation over the complex scientific issues, lower federal courts rebuffed agency preference for informal rule making and imposed procedural requirements that go beyond those of informal rule making but that fall short of the formal rule-making requirements under the Administrative Procedure Act. In other words, courts fashioned and imposed hybrid rule-making requirements.[117] Courts moreover have been inclined toward extension of the substantial-evidence test to informal rule making. The traditional distinction between judicial review under the arbitrary-and-capricious and the substantial-evidence standards has thereby been obscured. The distinction broke down only partially because of judicial intervention. Congress, in delegating extensive authority to agencies—under the Occupational Safety and Health Act, for example—provided for both informal rule making and close judicial scrutiny of regulations on the substantial-evidence standard.[118] Thus, as the Court of Appeals for the District of Columbia Circuit observed, "[There is an] emerging consensus of the Courts of Appeals that the distinction between the arbitrary and capricious standard and substantive evidence review is largely semantic.[119] Concomitantly, the traditional distinction between questions of fact and questions of law has been eroded.[120] Judicial intervention and intensive review of regulatory decisions were inevitable given both the broad legislative role and crucial policy choices delegated to federal agencies and the scientific uncertainties about the evidence adduced for health, safety, and environmental standards.

The form and extent of judicial intervention reflects differences in judicial self-perception. Federal judges, of course, differ in their perceptions of their role and capacity as generalists to scrutinize complex, science-policy regulations. The Court of Appeals for the District of Columbia Circuit, which reviews the vast majority of agency rule making, was sharply divided for a decade over the kind of review that should be given to the procedural and scientific basis for regulatory actions. The complexity of the issues is not the fundamental problem, Judge David Bazelon noted, since courts review no less vexing Federal Communications Commission (FCC) and SEC decisions. Rather, science-policy disputes confront judges, who have no special expertise to assess the merits of competing scientific arguments, with vexatious issues over which no scientific consensus exists. Even if agreement did exist within the scientific community, basic moral-political choices would still remain—choices that should be ventilated in a public forum. For Judge Bazelon, lack of judicial competence implies judicial self-restraint. Courts should not substitute their judgments for those of scientific and administrative experts: "Courts are not the agency either to resolve the factual disputes or to make the painful value choices."[121] Judges, however, are presumably experts on process, and adherence to judicial self-restraint on substantive matters need not preclude judicial activism on procedural grounds. According to Judge Bazelon:

> [I]n cases of great technological complexity, the best way for courts to guard against unreasonable or erroneous administrative decisions is not for judges themselves to scrutinize the technical merits of each decision. Rather, it is to establish a decisionmaking process which assures a reasoned decision that can be held up to the scrutiny of the scientific community and the public.[122]

The federal judiciary should supervise the process of administrative decision making and, when necessary, impose procedural requirements in order to ensure public participation, open discussion, and the reasoned elaboration of the scientific basis for regulations.

By contrast, the late Judge Harold Leventhal argued that although deference should be shown to administrative and scientific expertise, a reviewing court should engage in "enough steeping" in technical matters to permit informed, substantive review of administrative action.[123] With searching substantive review, a judge "becomes aware, especially from a combination of danger signals, that the agency has not really taken a 'hard look' at the salient problems, and has not genuinely engaged in reasoned decisionmaking."[124] The hard-look approach advanced by Judge Leventhal requires that judges do more than, as Judge Henry J. Friendly said, simply look for "good faith efforts of agencies."[125] When reviewing agency actions, for example, under the NEPA[126] or based on the FWPCA's requirement of technological feasibility,[127] a court should "penetrate to the underlying decisions of an agency to satisfy itself that the agency has exercised a reasoned discretion with reasons that do not deviate from or ignore the ascertainable legislative intent."[128] Although courts should not innovate procedurally or substitute more-rigorous processes for those adopted by agencies, judges are capable of informing themselves about scientific and technical matters and thus may scrutinize, criticize, and overturn the basis for administrative decisions.

The hard-look approach demands a great deal of work from judges and may provide a pretense for judicial activism rather than judicial self-restraint. Unfortunately, not all federal-court judges have either the intellectual ability or the personal dedication of Judge Leventhal. They typically find complex science-policy litigation perplexing, taxing, and time consuming, even though they may and do occasionally appoint experts to advise them.[129] As Judge Gibbons remarked:

> In environmental litigation we are constantly placed in a position of choosing between the lies told by the fisherman's expert and the lies told by utility companies' experts. The overwhelming temptation for an appellate court is to accept the original fact finder's conclusion as to which expert was telling the smallest lie.[130]

Judges are usually dependent on the resourcefulness of their law clerks who, for the most part, have no more training in scientific and technical matters than they.[131]

In *Vermont Yankee Nuclear Power Corp.* v. *Natural Resources Defense Council, Inc.,*[132] the Supreme Court endorsed the hard-look approach, repudiating the District of Columbia Court of Appeals's opinion written by Judge Bazelon.[133] Judge Bazelon had argued that "substantive review of mathematical and scientific evidence by technically illiterate judges is dangerously unreliable" and therefore that review should be confined to the procedural dimensions of agency rule making.[134] In *Vermont Yankee,* Justice William Rehnquist explicitly rejected Judge Bazelon's strict-procedures-ensure-correct-decisions approach as judicial "Monday-morning quarterbacking."[135] Formal adjudicatory procedures do not assure the correct or best substantive decision and might prove pernicious by encouraging agencies to adopt formal adjudicatory procedures simply to avoid judicial review and reversal of their regulatory actions. Justice Rehnquist also emphasized that such an activist exercise of judicial review "clearly runs the risk of 'propel[ling] the court into the domain which Congress has set aside exclusively for the administrative agency' " and thwarts the "very basic tenet of administrative law that agencies should be free to fashion their own rules of procedure."[136] Instead of dictating quasi-judicial procedures for agency resolution of complex science-policy disputes, lower federal courts should take a hard look at the substantive basis for agency decisions in order to ensure reasoned rule making.

The precise contours of the hard look approach and judicial demands for a reasoned elaboration of agency decisions remains difficult to ascertain and predict. In part, this is because the hard-look approach may vary with a judge's posture toward judicial self-restraint or activism. Moreover, the approach is problematic because courts must review agency decisions based on crosscutting, overlapping, and disparate legislative mandates for regulating purported toxins and carcinogens. Under some federal statutes—for example, the Consumer Product Safety Act;[137] the Federal Insecticide, Fungicide, and Rodenticide Act; [138] and the Toxic Substances Control Act[139]—agencies may balance the risks and benefits of carcinogens. Statutes requiring balancing, however, are by no means uniform. Some explicitly direct agencies to balance costs, risks, and benefits, whereas other statutes provide only vague mandates for balancing. In still other instances—for example, the Occupational Safety and Health Act[140] and provisions of the Clean Air[141] and Clean Water Acts[142]—the technological feasibility of a regulation must be considered. Finally, under the Delaney clause of the Food, Drug and Cosmetic Act,[143] no degree of carcinogenic risk is acceptable and no cost-benefit analysis of a substance, such as saccharin, is permitted.[144] Different legislative mandates for cost-risk-benefit analysis significantly affect regulatory action, the opportunities for court challenges, and therewith the scope of judicial review and intervention.

The Supreme Court's decisions in *Industrial Union Department, AFL-CIO* v. *American Petroleum Institute*[145] and *American Textile Manufacturers Institute, Inc.* v. *Donovan*[146] illustrate the problematic nature of the hard-look approach and the temptation for courts to demand more-reasoned elaboration of agency decisions than congressionally mandated. In *American Petroleum Institute,* the Court divided five-to-four in striking down OSHA's standard reducing the permissible exposure limit on airborne concentrations of benzene from 10 parts per million parts of air (10 ppm) to 1 part per million (1 ppm).[147] The Secretary of Labor had gathered more than fifty volumes of exhibits and testimony in deciding that exposure to benzene creates a risk of cancer, chromosomal damage, and a variety of nonmalignant but potentially fatal blood disorders, even at the level of 1 ppm. OSHA maintains that there is no safe-threshold level of carcinogenic exposure and that the present state of science renders it impossible to calculate the number of lives that would be saved by a 1 ppm benzene standard.[148] Justice John Paul Stevens, for a plurality, nevertheless found OSHA's rationale for lowering the permissible exposure limit from 10 ppm to 1 ppm unsupported. Justice Stevens, moreover, strongly implied that OSHA should undertake a cost-benefit analysis when their regulatory standards are based on technological- and economic-feasibility and carcinogenic-risk assessments.[149] Dissenting Justice Thurgood Marshall was thus prompted to respond: "[A]s the Constitution 'does not enact Mr. Herbert Spencer's *Social Statics,'* . . . so the responsibility to scrutinize federal administrative action does not authorize this Court to strike its own balance between the costs and benefits of occupational safety standards.''[150] A year later the Court addressed the confusion invited by Justice Stevens's plurality opinion. In *Donovan,* a bare majority upheld OSHA's cotton-dust standard, limiting occupational exposure to cotton dust.[151] Justice William Brennan, writing for the Court, ruled that cost-benefit analysis was not congressionally required under the Occupational Safety and Health Act, unlike other statutes,[152] and indicated that it would be improper for courts to require such an analysis in order to achieve more-reasoned administrative decision making.

The hard-look approach requires close judicial scrutiny of the scientific basis for health, safety, and environmental regulations. In adhering to the hard-look approach since *Vermont Yankee,* the District of Columbia Circuit Court of Appeals has appeared to adopt a moderate posture when reviewing complex science-policy issues. The court recognizes that ''a rule-making agency necessarily deals less with 'evidentiary' disputes than with normative conflicts, projections from imperfect data, experiments and simulations, educated predictions, differing assessments of possible risks, and the like.''[153] Although finding inadequate the evidence adduced by the EPA for its emission standard for lime manufacturers, the court emphasized the import of judicial self-restraint in substantive review of administrative regulations.[154] In subsequently upholding the EPA's promulgation of national ambient air-quality standards for lead[155] and ozone,[156] two different panels of the court of appeals underscored the propriety of judicial deference to agencies on matters of scientific and medical uncertainty.

The hard-look approach should not and need not lead to judicial second guessing—substitution of judicial for administrative judgments on complex science-policy disputes. Instead, the judiciary should modestly review and demand the reasoned elaboration of agency decision making. Reasoned administrative decision making requires that an agency clearly state its scientific and technological assumptions;[157] reveal the data, methodology, and literature upon which it relied;[158] explain its rejection of alternative theories and technological options;[159] and, finally, articulate its rationale for setting a particular regulatory standard in a manner that permits public comment and judicial review.[160]

The Forms and Limitations of Judicial Intervention

The judiciary has important regulatory functions and a participatory role in the political dynamics of resolving complex, polycentric science-policy disputes. At common law, when defining the nature and contours of legal rights, courts historically have resolved more or less complicated scientific and technological disputes. In balancing competing legal claims, judges render their own assessments of the utility of industrial and technological processes. When imposing legal doctrines such as strict product liability, judicial rulings have a regulatory impact—they structure private-sector decision making and require industries to internalize the social costs of their hazardous processes and consumer goods. The federal judiciary also emerged in the last decade as an important participant in regulatory politics. Federal courts have a crucial supervisory role in monitoring the decision-making process and scientific evidence adduced for health, safety, and environmental regulations. All major federal regulation appears ultimately to require the judicial imprimatur.

"Assessment of risk," Judge J. Skelly Wright notes, "is a normal part of judicial and administrative fact-finding."[161] Nonetheless, inherent problems remain with risk assessment and resolution of polycentric science-policy disputes in the judicial arena. Contrary to critics of regulatory agencies and advocates of reliance on the marketplace and judicially enforced property rights, the judiciary seems an inauspicious forum for public-policy formulation.[162] The adjudicatory process is an inappropriate mechanism for resolving science-policy disputes and undertaking technology assessments. Also, the impropriety cum illegitimacy of relying on the judiciary as the primary forum for resolving science-policy controversies is underscored by the limitations of judicial expertise and accountability.

As passive, reactive institutions that must await litigation that focuses retroactively on disputes, courts are neither structurally nor situationally predisposed to assess the impact and utility of industrial and technological developments. Judicial rulings, moreover, are inevitably limited, fragmentary, and incremental. They are also often redundant and conflicting given our dual constitutional system and fifty-state jurisdictions. The potential for discretionary injustice when imposing tort liability exists just as private-law adjudication appears inequitably to distribute compensation. Maldistribution of awards arises not only because of the problems of proving causation in health, safety, and environmental litigation but also simply because of the constraints of present legal doctrines, rules of evidence, and statutes of limitations. The judicial arena and private-law litigation have limited utility for assessing and forcing the internalization of the social costs of allegedly hazardous processes and goods that are thinly spread over dispersed and unorganized populations.

Adjudication fundamentally aims at procedural fairness, not fact finding, in imposing social sanctions. The accusatorial process in a number of ways frustrates judicial finding and resolution of science-policy disputes. More important, the adjudicatory process does not and cannot settle scientific controversies, which depend on achieving consensus within the scientific community itself. When rebuking an attack on the scientific basis for the EPA's regulation of gasoline additives, the District of Columbia Circuit Court of Appeals thus observed:

> Petitioners demand sole reliance on *scientific* facts, on evidence that reputable scientific techniques certify as certain. Typically, a scientist will not so certify evidence unless the probability of error, by standard statistical measurement, is less than 5%. That is, scientific fact is at least 95% certain.
>
> Such certainty has never characterized the judicial or the administrative process. It may be that the "beyond the reasonable doubt" standard of criminal law demands 95% certainty. . . . Since *Reserve Mining* was adjudicated in court, this standard applied to the court's fact-finding. Inherently, such a standard is flexible; inherently, it allows the fact-finder to assess risks, to measure probabilities, to make subjective judgments. Nonetheless, the ultimate finding will be treated, at law, as fact and will be affirmed if based on substantial evidence, or, if made by a judge, not clearly erroneous.[163]

The adjudicatory process was neither designed for nor proves amenable to resolving scientific uncertainties either within the judicial forum or when imported to the administrative arena and rule-making procedures. Private-law adjudication may settle local and limited disputes, but this reflects more the circumstances of the particular controversy than the merits of the adjudicatory process and judicial assessments of societal interests. Litigation of polycentric science-policy disputes engenders lengthy, costly, and not always conclusive trials and in turn precipitates out-of-court negotiations. Thus, the political import of adjudication of science-policy disputes centers not on providing a mechanism for settling scientific disagreements but instead serves as a catalyst for achieving political compromises over the normative conflicts attending the quest for scientific certainty and formulation of science policies.

The limited resources, training, and opportunities to obtain and competently use scientific and technical materials further bodes ill for reliance on the judicial arena to regulate hazardous industrial processes and consumer goods. Judicial assessments of risks are not merely likely to be less informed than those rendered in administrative arenas. Administrative agencies are frequently required by statute expressly to identify and in some fashion to balance risks, costs, and benefits, and their regulatory decisions remain subject to review by federal courts, Congress, and the executive branch.[164] By contrast, judges and juries need neither identify the values nor quantify the evidence adduced for their decisions. That judicial risk assessments remain relatively unstructured is evident from the Fifth Circuit Court of Appeals's explanation of judicial balancing in product-liability litigation:

> [T]he reasonable man standard becomes the fulcrum for a balancing process in which the utility of the product properly used is weighed against

whatever dangers of harm inhere in its introduction into commerce . . . [I]f the potential harmful effects of the product . . . outweigh the legitimate public interest in its availability, it will be declared unreasonable per se and the person placing it on the market held liable.[165]

Proposals for institutional changes and increased staff supports—for example, establishment of specialized courts, creation of a so-called science court, and appointment of masters or social- and natural-science clerks to help judges—might improve judicial competence and the capacity to assess complex scientific and technical issues and to thereupon balance societal interests.[166] Still—as always in a constitutional democracy—the issue of political accountability and legitimacy remains—that is, the auspiciousness and propriety of relying on judges, who are neither elected nor directly politically accountable, to render major regulatory decisions in the public interest.

The chilling effects of judicial independence, as much as the limitations of judicial competence, the institutional structure of the judicial forum, and the adjudicatory process, dictate a modest role for the judiciary when intervening in science-policy disputes and when reviewing legislative and administrative assessments of technology and the promulgation of health, safety, and environmental regulations. Technological and scientific advances are subject to diverse, overlapping, and at times, conflicting assessments in both the private and public sectors. Within the framework of the Constitution and a pluralistic political process, the Congress and the executive branch, not the judiciary, are the legitimate and primary institutions for assessing, structuring, and resolving the vexations science-policy issues posed by industrial and technological developments.

Notes

1. David Bazelon, "The Courts and the Public: Policy Decisions about High Technology and Risk" (Speech delivered at the University of Pennsylvania, 4 April 1981), pp. 3–4. The author expresses his appreciation to Judge Bazelon for making a copy of his speech available.

2. Ibid. See also Bazelon, "Coping with Technology through the Legal Process," *Cornell Law Review* 62 (1977):817; and compare Harold Leventhal, "Environmental Decisionmaking and the Role of the Courts," *University of Pennsylvania Law Review* 122 (1974):509; J. Skelly Wright, "New Judicial Requisites for Informal Rulemaking: Implications for the Environmental Impact Statement," *Administrative Law Review* 29 (1977): 59; Carl Mc Gowan, "Congress, Court, and Control of Delegated Power," *Columbia Law Review* 77 (1977):1119; and Henry J. Friendly, "The Courts and Social Policy," *University of Miami Law Review* 33 (1978):21.

3. See, for example, *Whalen* v. *Roe,* 97 S.Ct. 869 (1977).

4. See, for example, *Diamond* v. *Chakrabarty,* 65 L.Ed.2d 144 (1980).

5. See, for example, *American Textile Manufacturers Institute, Inc.* v. *Donovan,* 101 S.Ct. 2478 (1981).

6. See, for example, *Ethyl Corporation* v. *Environmental Protection Agency,* 541 F.2d 1 (1976); *International Harvester Co.* v. *Ruckelshaus,* 478 F.2d 615 (1973); and *Hercules Incorporated* v. *Environmental Protection Agency,* 598 F.2d 91 (1978).

7. See, for example, *Environmental Defense Fund, Inc.* v. *Ruckelshaus,* 439 F.2d 584 (1971).

8. See, for example, *Environmental Defense Fund* v. *Environmental Protection Agency,* 636 F.2d 1267 (1980); and *Environmental Defense Fund* v. *Environmental Protection Agency,* 598 F.2d 62 (1978).

9. See, for example, *Pactra Industries, Inc.* v. *Consumer Product Safety Commission,* 555 F.2d 677 (1977); and *Society of Plastics Industry, Inc.* v. *Occupational Safety and Health Administration* 509 F.2d 1301 (1975).

10. See, for example, *Industrial Union Department, AFL-CIO* v. *American Petroleum Institute,* 100 S.Ct. 2844 (1980).

11. See, for example, *Sindell* v. *Abbott Laboratories,* 26 Cal.3d 588, 607 P.2d 924 (1980), *cert. denied.* 101 S.Ct. 286 (1980).

12. See, for example, *United States* v. *Reserve Mining Company,* 380 F. Supp. 11 (1974).

13. See, for example, *Vermont Yankee Nuclear Power Company* v. *Natural Resources Defense Council, Inc.,* 435 U.S. 519 (1978).

14. See Alexis de Tocqueville, *Democracy in America* (New York: Vantage Books, 1965), p. 156; Maurice Rosenberg, "Let's Everybody Litigate," *Texas Law Review* 50 (1972):1349; Laurence Silberman, "Will Lawyering Strangle Democratic Capitalism?" *Regulation* (1978):15; and Bayless Manning, "Hyperplexis: Our National Disease," *Northwestern University Law Review* 71 (1977):767.

15. See Jethro Lieberman, *The Litigious Society* (New York: Basic Books, 1981). p. 192, n. 14; and Director of the Administrative Office of the United States Courts, *Management Statistics for United States Courts 1980* (Washington, D.C., 1981), pp.13 and 129.

16. J. Woodford Howard, Jr., *Courts of Appeals in the Federal Judicial System* (Princeton, N.J.: Princeton University Press, 1981), p. 13.

17. Milton Katz, "Decision-making in the Production of Power," *Scientific American* 223 (1971):191, 198.

18. See, for example, Robert Field, "Statutory and Institutional Trends in Governmental Risk Management: The Emergence of a New Structure" (Paper prepared for the Committee on Risk and Decision Making,

National Academy of Sciences, (1981); and Environmental Law Institute, *An Analysis of Past Federal Efforts to Control Toxic Substances* (Washington, D.C.: Council on Environmental Quality, 1978).

19. Council on Environmental Quality, *Tenth Annual Report of the Council on Environmental Quality* (Washington, D.C.: Government Printing Office, 1979).

20. See *American Textile Manufacturers Institute, Inc.* v. *Donovan,* 101 S.Ct. 2478 (1981).

21. See, for example, Petition of the Environmental Defense Fund to the Consumer Product Safety Commission, 26 March 1976; and Official Transcripts of Proceedings before the Consumer Product Safety Commission, in the Matter of Commission Meeting on Petition of TRIS-Treated Sleepwear, 8 March 1977 and 4, 22, and 31 April 1977. The author appreciates the efforts of Ms. Pricilla Martinez, of the Consumer Product Safety Commission, in making these and other materials available.

22. See Leventhal, "Environmental Decisionmaking."

23. *Industrial Union Department* v. *Hodgson,* 499 F.2d 467, 474 (D.C. Cir. 1974); and *Ethyl Corporation* v. *Environmental Protection Agency,* 541 F.2d 1, 26–27, 47, n. 97 (1976).

24. See, generally, Donald Horowitz, *The Courts and Social Policy* (Washington, D.C.: The Brookings Institution, 1977); and David M. O'Brien, "The Seduction of the Judiciary: Social Science and the Courts," *Judicature* 64 (1980):8, 13–16.

25. Transcript of Proceedings, *United States* v. *Reserve Mining Company,* U.S. District Court, District of Minnesota, Fifth Division, No. 5-72, Civil 19, 14 November 1975, pp. 21–22. My appreciation to Mr. O.E. Breviu for expeditiously obtaining a copy of the transcript for my use.

26. Ibid., pp. 4–5.

27. *Reserve Mining Company* v. *Lord,* 529 F.2d 185–186 (1976).

28. See, Michael Polanyi, *The Logic of Liberty: Reflections and Rejoinders* (Chicago: University of Chicago Press, 1951), p. 171; and Lon Fuller, "The Forms and Limits of Adjudication," *Harvard Law Review* 92 (1978):353.

29. Leon Green, "Tort Law: Public Law in Disguise," *Texas Law Review* 38 (1959–1960):1, 269. See also Milton Katz, "The Function of Tort Liability in Technology Assessment," *Cincinnati Law Review* 38 (1969): 587.

30. Ibid., p. 257.

31. See, generally, U.S. Congress, Senate, Committee on Environment and Public Works, *Six Case Studies of Compensation for Toxic Substances Pollution: Alabama, California, Michigan, Missouri, New Jersey, and Texas, A Report of the Congressional Research Service* 96th Cong., 2d sess., 1980.

32. On the logic of collective action and the problem of freeriders, see, generally, Mancur Olsen, *The Logic of Collective Action* (Chicago: University of Chicago Press, 1965).

33. In recent years, Congress has passed special legislation to compensate individuals exposed to particular toxic and carcinogenic substances. See the Comprehensive Environmental Response, Compensation and Liability Act, 42 U.S.C. §§9601–9657 (1980) (permitting compensation for victims of hazardous-waste disposal); and Black Lung Benefits Reform Act, 30 U.S.C. §901 (permitting compensation for coal miners contracting pneumoconiosis).

34. See Arthur Pigou, *The Economics of Welfare,* 4th ed. (London: MacMillan & Co., 1932); R.H. Coase, "The Problems of Social Cost," *Journal of Law and Economics* 3 (1960):1; A.R. Pelst and R. Turvey, "Cost-Benefit Analysis: A Survey," *The Economics Journal* (1965):683; James Buchanan and William Stubblebine, "Externality," *Economica* 29 (1963):371; Ralph Turvey, "On Divergences between Social Cost and Private Cost," *Economica* 30 (1963):309; and Robert Haveman and Burton A. Weisbrod, "Defining Benefits of Public Programs: Some Guidance for Policy Analysts," *Policy Analysis 1* (1975):169.

35. See Burton Weisbrod, *Public Interest Law: An Economic and Institutional Analysis* (Berkeley: University of California Press, 1978), pp. 44, 46, 88–89, 96–97.

36. *Sindell* v. *Abbott Laboratories*, 26 Cal.3d 588, 607 P.2d 924 (1980), *cert. denied* 101 S.Ct. 286 (1980).

37. See *Washington Insurance Newsletter* 31 (22 October 1980):2.

38. Laurence Tribe, *Channeling Technology Through Law* (Chicago: Bracton Press, 1973), p. 56.

39. See, for example, *Arvidson* v. *Reynolds Metals Co,* 125 F.Supp. 481 (W.D. Wash. 1954) (fluorine gas and minute particles deposited on and over plaintiff's land not a trespass); and *Batten* v. *United States,* 125 F.2d 580 (10th Cir. 1962) (no recovery for airport noise unless plane flies directly over plaintiff's property); but see *Reynolds Metals Co.* v. *Martin,* 337 F.2d 780 (1964) (fluorine fumes constitute both trespass and nuisance).

40. Lieberman, *The Litigious Society,* p. 19.

41. See William Prosser, "Private Action for Public Nuisance," *Virginia Law Review* 52 (1966):997.

42. See, "Pollution Control" *American Jurisprudence* 2d. §149, 61.

43. See, for example, *Commonwealth* v. *Barnes and Trucker Co.* (Sup. Ct. Pa. 1974) 4 E.L.R. 20545 (1974).

44. See, for example, *Green* v. *Castle Concrete Co.,* (Dis. Ct. El Paso, County 1971 2 E.L.R. 20347 (1971).

45. See *Village of Wilsonville* v. *SCA Services, Inc.,* 396 N.E. 2d 552 (Ill. App. Ct. 1979).

46. *Illinois* v. *City of Milwaukee,* 406 U.S. 91 (1972).

47. Federal Water Pollution Control Act, as amended by the Clean Water Act, 33 U.S.C. §§1251–1376 (1972).

48. *Township of Long Beach* v. *City of New York,* 445 F.Supp. 1203 (D.N.J. 1978).

49. *City of Evansville* v. *Kentucky Liquid Recycling, Inc.,* 604 F.2d 1008 (7th Cir. 1979).

50. See Comment, "Hazardous Waste: EPA, Justice Invoke Emergency Authority, Common Law in Litigation Campaign against Dump Sites," 10 E.L.R. 10034 (1980).

51. See, generally, Horowitz, *Courts and Social Policy;* and O'Brien, "Seduction of the Judiciary."

52. *City of Milwaukee* v. *Illinois,* 11 E.L.R. 20406 (1980).

53. See *Erie Railroad Co.* v. *Tompkins,* 304 U.S. 64, 78 (1938); and *United States* v. *Hudson & Goodwin,* 11 U.S. (7 Cranch) 32 (1812).

54. *City of Milwaukee* v. *Illinois,* 11 E.L.R. 20406, 20414 (1980) (Blackmun, J., dis. op.).

55. *Rylands* v. *Fletcher,* 3 Hurl. C. 774 (1865), L.R. 1 Ex. 265 (1866), L.R. 3H.L. 330 (1868).

56. *Rylands* v. *Fletcher,* L.R. 1 Ex. 265, 279 (1866).

57. *Restatement (Second) of Torts* §§519–520 (Tentative Draft No. 10, 1964).

58. *Holman* v. *Athens Empire Laundry Co.* 149 Ca. 345, 100 S.E. 207 (1919).

59. *Luthinger* v. *Moore,* 31 Cal.2d 489, 190 P. 2d 1 (1948).

60. *Gotreaux* v. *Gary,* 232 La. 373, 94 So.2d 293 (1957).

61. *United States* v. *FMC Corporation,* 572 F.2d 902 (1978).

62. *United States* v. *FMC Corporation,* 572 F.2d 902 (1978).

63. See, for example, Paul Rheingold, Norman Landau, and Michael Canavan, eds., *Toxic Torts: Tort Actions for Cancer and Lung Disease Due to Environmental Pollution* (Washington, D.C.: Association of Trial Lawyers of America, 1977); Richard Posner, "A Theory of Negligence," *Journal of Legal Studies* 2 (1973):205; Guido Calabresi, "Some Thoughts on Risk Distribution and the Law of Torts," *Yale Law Journal* 70 (1961):499; and George Fletcher, "Fairness and Utility in Tort Theory," *Harvard Law Review* 85 (1972):537.

64. See Joyce Egginton, *The Poisoning of Michigan* (New York: W.W. Norton, 1980); and Edwin Chen, *PBB: An American Tragedy* (Englewood Cliffs, N.J.: Prentice-Hall, 1979).

65. See Ellen E. Grezech, "PBB," In *Who's Poisoning America,* Ralph Nader, Ronald Brownstein, and John Richard, eds. (San Fransciso: Sierra Club Books, 1981), pp. 60, 63.

66. See, for example, *Martinez* v. *Dixie Carriers, Inc.,* 529 F.2d 457 (1976).

67. *Boyl* v. *California Chemical Company,* 221 F.Supp. 669 (1963).

68. Ibid., p. 676.

69. *(Second) Restatement of the Law of Torts* §291 (Tentative Draft, 2d ed., 1964).

70. See, for example, Richard Epstein, "A Theory of Strict Liability," *Journal of Legal Studies* 2 (1973):29; Richard Posner, "Strict Liability: A Comment," *Journal of Legal Studies* 2 (1973):205; Guido Calabresi and Jon Hirshoff, "Toward a Test for Strict Liability in Torts," *Yale Law Journal* 81 (1972):1055; and Richard Epstein, "Products Liability: The Gathering Storm," *Regulation* (1977):15.

71. See, for example, *Borel* v. *Fibreboard Paper Products Corporation* 493 F.2d 1076 (1973).

72. *In re "Agent Orange" Product Liability Litigation,* 506 F.Supp. 737 (1979); and *In re "Agent Orange" Product Liability Litigation,* 635 F.Supp. 987 (1980). See also John Rabinovich, "The Politics of Poison," in *Who's Poisoning America,* Ralph Nader, Ronald Brownstein, and John Richard, eds. (San Francisco: Sierra Club Books, 1981), p. 240.

73. Model Uniform Product Liability Act, §104(a), *Federal Register* 44 (1979):62721.

74. Ibid., 5104(b).

75. *Greenman* v. *Yuba Power Products, Inc.,* 59 Cal.2d 57, 63–64, 377 P.2d 897, 901 (1963).

76. *Helene Curtis Industries, Inc.* v. *Pruitt,* 385 F.2d 841, 849 (1967).

77. Ibid.

78. According to Leonard Andrews, publisher of *Asbestos Litigation Reporter,* and Richard Lippes, attorney for Love Canal residents, as reported in Council on Environmental Quality, *Environmental Quality-1980,* Eleventh Annual Report (Washington, D.C.: Government Printing Office, 1981), p. 247, ns. 112 and 113.

79. *Silkwood* v. *Kerr-McGee Corporation,* 485 F.Supp. 566 (D.C. Okla. 1979). On 13 December 1981, a federal court of appeals reverted the $10.5 million award. See "$10.5 Million Award Is Upset," *The New York Times,* 13 December 1981, p. A42.

80. See "The Devils in the Product Liability Laws," *Business Week*, 12 February 1979, p. 72.

81. Lieberman, *The Litigious Society,* p. 49. See also Jeffrey O'Connell, *The Lawsuit Lottery* (New York: Free Press, 1979).

82. U.S. Congress, House, Committee on Small Business, *Product Liability Insurance,* 95th Cong., 2d sess., H.R. No. 95-97, p. 6 (1978). See Lieberman, *The Litigious Society,* pp. 48–52.

83. Lieberman, *The Litigious Society,* p. 48, quoting California Citizen's Commission on Tort Reform, *Righting the Liability Balance,* p. 111 (San Francisco, Calif.: Citizen Commission on Tort Reform, 1977).

84. See U.S. Congress, Senate, *Six Case Studies.*

85. See, generally, Abraham Chayes, "The Role of the Judge in Public Law Litigation," *Harvard Law Review* 89 (1976):1281.

86. *Karjala* v. *Johns-Manville Products Corporation,* 523 F.2d 155, 160 (9th Cir.,1975).

87. See, generally, Martin Shapiro, "Toward a Theory of Stare Decisis," *Journal of Legal Studies* 1 (1972):125.

88. Harold Green, *The New Technological Era: A View from the Law,* Monograph 1 (Washington, D.C.: George Washington University, Program of Policy Studies, 1967). See also Green, "The Role of Law and Lawyers in Technology Assessment," *Atomic Energy Law Journal* 13 (1971):246.

89. See *Assessment of Technologies for Determining Cancer Risks from the Environment* (Washington, D.C.: Government Printing Office, 1981) pp. 31–65, 113–175; National Academy of Sciences, *Principles for Evaluating Chemicals in the Enviroment* (Washington, D.C., 1975); and Howard Hiatt, James Watson, and Jay Winston, eds., *The Origins of Human Cancer,* 3 vols. (Cold Spring Harbor, N.Y.: Cold Springs Laboratory, 1977).

90. For a further discussion, see Robert Harley, "Proof of Causation in Environmental Litigation," in *Toxic Torts: Tort Actions for Cancer and Lung Disease Due to Environmental Pollution,* Paul Rheingold, Norman Landau, and Michael Canavan, eds., p. 403. (Washington, D.C.: Association of Trial Lawyers of America, 1977).

91. For a further discussion, see Barry Boyer, "Alternatives to Administrative Trial-Type Hearings to Resolve Complex Scientific, Economic, and Social Issues," *Michigan Law Review* 71 (1972):111; and Jeffrey Martin, "Procedures for Decision Making under Conditions of Scientific Uncertainty: The Science Court Proposal," *Harvard Journal of Legislation* 16 (1979):443.

92. Statistical and medical evidence and testimony is often precluded from introduction at trial or treated as hearsay evidence. See *Federal Rules of Evidence,* Rule 703; and *Hazelwood School District* v. *United States,* 433 U.S. 277 (1977) (statistics in use of employment-discrimination litigation). See also Richard Phelan, "Proof of Cancer from a Legal Viewpoint," in *Toxic Substances: Problems in Litigation,* S. Birnbaum, ed. (New York: Practicing Law Institute, 1981), p. 155.

93. See *Urie* v. *Wisconsin,* 337 U.S. 163 (1949) (statute of limitations does not commence until accumulated effects manifest themselves—silicosis case); but also see *Karjala* v. *Johns-Manville Products Corp.,* 523 F.2d 155 (9th Cir. 1975).

94. See, for example, *(Second) Restatement of Torts,* §496A; *Thomas* v. *Kaiser Agricultural Chemicals,* 81 Ill.2d 206, 407 N.E.2d 32 (1980) (assumption of risk); *Woodhill* v. *Parke Davis & Company,* 37 Ill.2d 304

402 N.E.2d 194 (1980); and *Dalke* v. *Upjohn Co.,* 555 F.2d 245, 248 (1977) (liability limited by state of scientific knowledge).

95. Julian Juergensmeyer, "Control of Air Pollution through the Assertion of Private Rights," *Duke Law Journal* (1967):1126, 1154. See also Paul Rheingold, "Civil Cause of Action for Lung Damage Due to Pollution of Urban Atmosphere," *Brooklyn Law Review* 33 (1966):17.

96. For a further discussion, see U.S. Congress, Senate, *Six Case Studies.*

97. *Union Electric Co.* v. *Environmental Protection Agency,* 427 U.S. 246, 257 (1976). See also, *Train* v. *Natural Resources Defense Council, Inc.,* 421 U.S. 60 (1975).

98. See, Bruce A. Ackerman and William T. Hassler, *Clean Coal/Dirty Air* (New Haven: Yale University Press, 1981), pp. 1–12; James O. Freedman, *Crisis and Legitimacy* (Cambridge: Cambridge University Press, 1978), pp. 31–89.

99. See, for example, *FCC* v. *Pottsville Broadcasting Co.,* 309 U.S. 134, 138–44 (1940).

100. *International Harvester Company* v. *Ruckelshaus,* 478 F.2n 615, 651 (1973).

101. See Stephen Breyer and Richard Stewart, *Administrative Law and Regulatory Policy* (Boston: Little, Brown & Co., 1979), pp. 922–925. See generally, Stewart, "The Reformation of American Administrative Law." *Harvard Law Review* 88 (1975):1667.

102. Administrative Procedure Act, 5 U.S.C.§706.

103. See, for example, *Association of Data Processing* v. *Camp,* 397 U.S. 150 (1970); *Scenic Hudson Preservation Conference* v. *FPC,* 354 F.2d 608 (1965), *cert. denied,* 384 U.S. 941 (1966); *Schlesinger* v. *Reservists Committee to Stop the War,* 418 U.S. 208 (1974); *United States* v. *Richardson,* 418 U.S. 166 (1974); and *Duke Power Co.* v. *Carolina Environmental Study Group,* 98 S.Ct. 2620 (1978).

104. See, for example, Resource Conservation and Recovery Act, 42 U.S.C. Section 1001.

105. See, for example, *Cannon* v. *University of Chicago,* 99 S.Ct. 1946 (1979); *City of Evansville* v. *Kentucky Liquid Recycling,* 604 F.2d 1008 (1979) (denial of private cause of action under Rivers and Harbor Act, 33 U.S.C. §13, in challenge of dumping of toxic chemicals into rivers).

106. See, for example, *Frothingham* v. *Mellon,* 262 U.S. 447 (1923); and compare *Flast* v. *Cohen,* 392 U.S. 83 (1969). See also Louis Jaffe, "The Citizen as Litigant in Public Actions: The Non-Hohfeldian or Ideological Plaintiff," *University of Pennsylvania Law Review* 116 (1968):1033.

107. *Association of Data Processing, Serria Organizations, Inc.* v. *Camp,* 397 U.S. 150, 153–154 (1970).

108. *United States* v. *Students Challenging Regulatory Agency Procedures,* 412 U.S. 669 (1973).

109. Ibid, at 686.

110. *Warth* v. *Seldin,* 422 U.S. 490 (1975); *Arlington Heights* v. *Metropolitan Housing Development Corp.,* 429 U.S. 252, 261 (1971); *Simon* v. *Eastern Kentucky Welfare Rights Organization,* 426 U.S. 26, 41–42 (1976); and *Duke Power Co.* v. *Carolina Environmental Study Group, Inc.,* 98 S.Ct. 2620 (1978).

111. See, for example, *Environmental Defense Fund* v. *Hardin,* 325 F.Supp. 1401 (1971) (attacking secretary of agriculture's sanction of use of Mirex to control of fire ants under the Federal Insecticide, Fungicide and Rodenticide Act).

112. See, for example, *Weinberger* v. *Hynson, Westcott and Dunning,* 412 U.S. 609 (1973) (upholding FDA's refusal of a hearing to a petitioner challenging withdrawal of a new drug application on efficacy grounds).

113. See, for example, *Public Citizen* v. *Foreman,* 631 F.2d 969 (1980) (challenge of FDA's ruling on nitrite).

114. See, for example, *Scenic Hudson Preservation Conference* v. *Federal Power Commission,* 354 F.2d 608 (1965) (challenging agency action for failing to support decisions with relevant data and permit public access to records). See also the discussion in the next section of the hard-look approach to judicial review of administrative rule making.

115. Administrative Procedure Act, 5 U.S.C. §706. See *Natural Resources Defense Council, Inc.* v. *Securities and Exchange Commission,* 423 F.Supp. 1190 (1977); *Dunlop* v. *Bachowski,* 421 U.S. 560 (1975); *Environmental Defense Fund, Inc.* v. *Blum,* 100 S.Ct. 1889 (1980) (arbitrary and capricious standard); *Weinberger* v. *Hynson, Westcott and Dunning,* 412 U.S. 609 (1973); *Citizens to Preserve Overton Park, Inc.* v. *Volpe,* 401 U.S. 402 (1971); and *Universal Camera Corp.* v. *NLRB,* 340 U.S. 474 (1951) (substantial-evidence standard).

116. See, generally, American Bar Association, *Federal Regulation: Roads to Reform* (Washington, D.C., 1979); Commission of Federal Paperwork, *Rulemaking* (Washington, D.C.: Government Printing Office, 1977); Carl A. Averbach, "Informal Rule Making," *Northwestern University Law Review* 72 (1977):15; Charles Ames and Steven McCracken, "Framing Regulatory Standards to Avoid Formal Adjudication: The FDA as a Case Study," *California Law Review* 64 (1976):14; Nathaniel Nathanson, "Probing the Mind of the Administrator," *Columbia Law Review* 75 (1975):721; William Pedersen, "Formal Records and Informal Rulemaking," *Yale Law Journal* 85 (1975):38; Robert Hamilton, "Procedures for the Adoption of Rules of General Applicability," *California Law Review* 60 (1972):1276; Paul Verkuil, "Judicial Review of Informal Rulemaking," *Virginia Law Review* 60 (1974):185; and David Shapiro, "The Choice of Rulemaking or Adjudication in the Development of Administrative Policy," *Harvard Law Review* 78 (1965):1601.

117. See, for example, *Environmental Defense Fund* v. *Environmental*

Protection Agency, 598 F.2d 62 (1978) (regulation of PCBs); *Aqua Slide 'n' Dive Corp.* v. *Consumer Product Safety Commission,* 569 F.2d 831 (1978) (regulation of swimming-pool safety); *Wellford* v. *Ruckelshaus,* 439 F.2d 598 (1971) (registration of herbicide 2,4,5-T); *United States* v. *Weeks,* 487 F.2d 342 (1973); and *Synthetic Organic Chemical Manufacturers Association* v. *Brennan,* 503 F.2d 1155 (1974) (review of OSHA rule making). See also Stephen Williams, "'Hybrid Rulemaking' under the Administrative Procedure Act: A Legal and Empirical Analysis," *University of Chicago Law Review* 42 (1975):401; and James DeLong, "Informal Rulemaking and the Integration of Law and Policy," *Virginia Law Review* 65 (1979):257.

118. Occupational Safety and Health Act, 29 U.S.C.A. §655(a,b,e,f).

119. *Pacific Legal Foundation* v. *Department of Transportation* 593 F.2d 1338, 1343, n. 35 (1979).

120. See, for example, *Environmental Defense Fund* v. *Environmental Protection Agency,* 598 F.2d 62 (1978); *American Iron and Steel Institute* v. *Occupational Safety and Health Administration,* 577 F.2d 825 (1978). See also Kenneth Culp Davis, "Facts in Lawmaking," *Columbia Law Review* 80 (1980):931; Louis Jaffe, "Judicial Review: Question of Law," *Harvard Law Review* 69 (1955):239; and Jaffe, "Judicial Review: Question of Fact," *Harvard Law Review* 69 (1956):1020.

121. Bazelon, "Courts and the Public," p. 822.

122. *International Harvester Co.* v. *Ruckelshaus,* 478 F.2d 615, 653 (1973) (Bazelon, J., con. op.). See also *Ethyl Corporation* v. *Environmental Protection Agency,* 541 F.2d 1, 66 (1976) (Bazelon, J., con. op.); *Wellford* v. *Ruckelshaus,* 439 F.2d 598 (1971); and *Natural Resources Defense Council, Inc.* v. *Nuclear Regulatory Commission,* 547 F.2d 633 (1976).

123. On Judge Leventhal's judicial philosophy, see Leventhal, "Environmental Decisionmaking;" Leventhal, "Principled Fairness and Regulatory Urgency," *Case Western Reserve Law Review* 25 (1974):66; and Samuel Estreicher, "Pragmatic Justice: The Contributions of Harold Leventhal to Administrative Law," *Columbia Law Review* 80 (1980):894.

124. See Judge Leventhal's opinions in *Greater Boston Television Corporation* v. *FCC,* 444 F.2d 841 (1970), *cert. denied,* 403 U.S. 923 (1971); and *Ethyl Corporation* v. *Environmental Protection Agency,* 541 F.2d 1 (1976).

125. *City of New York* v. *United States,* 344 F.Supp. 929 (E.D.N.Y. 1072). See also Henry J. Friendly, "Some Kind of Hearing," *University of Pennsylvania Law Review* 123 (1975):1267; and Friendly, "The Courts and Social Policy," *University of Miami Law Review* 33 (1978):21.

126. National Environmental Policy Act of 1969, 42 U.S.C.A. §4321. See, for example, *International Harvester Co.* v. *Ruckelshaus,* 478 F.2d 615 (1973).

127. Federal Water Pollution Control Act Amendments of 1972, 33

U.S.C.A. §1317(a)(2,4,5). See for example, *Hercules, Inc.* v. *Environmental Protection Agency,* 598 F.2d 91 (1978).

128. *Greater Boston Television Corporation* v. *FCC,* 444 F.2d 841, 850 (1970).

129. See, for example, Federal Rules of Civil Procedure, Rule 53 (appointment of masters). Courts have used scientific experts in certain circumstances; see *Reserve Mining Company* v. *United States,* 498 F.2d 1073 (1974); and discussion in text at note 25.

130. Quoted by Maurice Rosenberg, "Contemporary Litigation in the United States," in *Legal Institutions Today,* Harry Jones, ed. (Chicago: American Bar Association, 1977), p. 158.

131. This statement is based on conversations with former clerks for judges of the District of Columbia Circuit Court of Appeals.

132. *Vermont Yankee Nuclear Power Corporation* v. *Natural Resources Defense Council,* 435 U.S. 519 (1978). See Antonin Scalia, "Vermont Yankee: The APA, the D.C. Circuit and the Supreme Court," in *The Supreme Court Review,* Philip Kurland, ed. (Chicago: University of Chicago Press, 1979), p. 345; William Rodgers, Jr., "A Hard Look at *Vermont Yankee:* Environmental Law under Close Scrutiny," *Georgetown Law Journal* 67 (1979):699; Richard Stewart, "*Vermont Yankee* and the Evolution of Administrative Procedure," *Harvard Law Review* 91 (1979): 1804; and Joel Yellin, "High Technology and the Courts: Nuclear Power and the Need for Institutional Reform," *Harvard Law Review* 94 (1981): 489.

133. *Natural Resources Defense Council, Inc.* v. *Nuclear Regulatory Commission,* 547 F.2d 633 (1976).

134. *Ethyl Corporation* v. *Environmental Protection Agency* 541 F.2d 1, 67 (1976).

135. *Vermont Yankee Nuclear Power Corporation* v. *Natural Resources Defense Council,* 435 U.S. 519, 547 (1978).

136. Ibid.

137. Consumer Product Safety Act, 15 U.S.C. §2064.

138. Federal Insecticide, Fungicide, and Rodenticide Act, 7 U.S.C. §§135–135(K).

139. Toxic Substances Control Act, 15 U.S.C. §§2601–2609.

140. Occupational Safety and Health Act, 29 U.S.C. §§651–678.

141. Clean Air Act, as amended, 42 U.S.C. §§7401–7642.

142. Clean Water Act, amendments to Federal Water Pollution Control Act, 33 U.S.C. §§1251–1376.

143. Food, Drug and Cosmetic Act, as amended, 21 U.S.C. §§301, 348 (food additives).

144. See National Research Council, *Safety of Saccharin and Sodium Saccharin in the Human Diet* (Washington, D.C.: National Academy of

Sciences, 1974). Congress extended its moratorium on the FDA's ban of saccharin. See U.S. Congress, Senate, "Saccharin Study and Labeling Act Amendments of 1981," 97th Cong., 1st sess., Rept. No. 97-140, 198.

145. *Industrial Union Department, AFL-CIO* v. *American Petroleum Institute,* 100 S.Ct. 2844 (1980). See also William Rodgers, Jr., "Judicial Review of Risk Assessments: The Role of Decision Theory in Unscrambling the Benzene Decision," *Environmental Law* 11 (1981):301.

146. *American Textile Manufacturers Institute, Inc.* v. *Donovan,* 101 S.Ct. 2478 (1981).

147. "Benzene Standard," *Federal Register* 43 (10 February 1978): 5918 as amended, *Federal Register* 43 (27 June 1978):27962.

148. See OSHA, "Proposed Rule on the Identification, Classification, and Regulation of Toxic Substances Posing a Potential Carcinogenic Risk," *Federal Register* 42 (4 October 1977):54148, 54165-54167; and "Benzene Standard."

149. *Industrial Union Department, AFL-CIO* v. *American Petroleum Institute,* 100 S.Ct. 2844, 2862-2863 (1980).

150. Ibid., at 2905 (Marshall, J., dis. op.).

151. "Cotton Dust Standard," *Federal Register* 43 (1978):27352. See also Part II, "Cotton Dust," *The Scientific Basis of Health and Safety Regulation,* Robert W. Crandall and Lester B. Lave, eds. (Washington, D.C.: The Brookings Institution, 1981), pp.71-116; and U.S., Department of Labor, *Report to Congress: Cotton Dust—Review of Alternative Technical Standards and Control Technologies* (14 May 1979).

152. See, for example, Energy Policy and Conservation Act of 1975, 42 U.S.C. §6295(c,d); Federal Water Pollution Control Act Amendments of 1972, 33 U.S.C. §1312(b)(1), (2); §1314(b)(1)(B); Clean Water Act Amendments of 1977, 33 U.S.C. §1314(b)(4)(B); and Clean Air Act Amendments of 1970, 42 U.S.C. §7545(c)(2)(B). See also *Environmental Protection Agency* v. *National Crushed Stone Assn.,* 101 S.Ct. 295 (1980) (best available technology).

153. *Amoco Oil Co.* v. *Environmental Protection Agency,* 501 F.2d 722, 735 (1974).

154. *National Lime Association* v. *Environmental Protection Agency,* 627 F.2d 416 10 E.L.R. 20366 (1980).

155. *Lead Industries Association, Inc.* v. *Environmental Protection Agency,* 10 E.L.R. 20643 (1980).

156. *American Petroleum Institute* v. *Costle,* 11 E.L.R. 20916 (1981).

157. See, for example, *International Harvester Co.* v. *Ruckelshaus,* 478 F.2d 615 (1973); and *Portland Cement Ass'n.* v. *Ruckelshaus,* 486 F.2d 735 (1973), *cert. denied,* 417 U.S. 921 (1974).

158. See, for example, *Portland Cement Ass'n.* v. *Ruckelshaus,* 486 F.2d 375, 393, 400 (1973).

159. See, for example, *International Harvester Co.* v. *Ruckelshaus,* 478 F.2d 615, 651 (1973) (Bazelon, J., con. op.); and *Amoco Oil Co.* v. *Environmental Protection Agency,* 501 F.2d 722, 738–739 (1974).

160. See, for example, *International Harvester Co.* v. *Ruckelshaus,* 478 F.2d 615, 648 (1973); *Kennecott Copper Corp.* v. *Environmental Protection Agency,* 462 F.2d 846, 849–850 (1972); *Environmental Defense Fund, Inc.* v. *Environmental Protection Agency,* 465 F.2d 528, 529 (1972); and *National Lime Association* v. *Environmental Protection Agency,* 10 E.L.R. 20366, 20384–20386 (1980).

161. *Ethyl Corporation* v. *Environmental Protection Agency,* 541 F.2d 1, 28, n. 58 (1976).

162. See, for example, Roger Meiners, "What to Do about Hazardous Products," in *Instead of Regulation,* Robert Poole, Jr., ed. (Lexington, Mass.: Lexington Books, D.C. Heath and Company, 1982), p. 285. See also Allen Kneese and Charles Schultze, *Pollution, Prices, and Public Policy* (Washington, D.C.: The Brookings Institution, 1975); and Charles Schultze, *The Public Use of Private Interest* (Washington, D.C.: The Brookings Institution, 1977).

163. *Ethyl Corporation* v. *Environmental Protection Agency,* 541 F.2d 1, 28, n. 58 (1976) (emphasis in original and citations omitted).

164. See text and notes 97–100, 115–136, and 144. See also U.S. Congress, House, *Presidential Control of Agency Rulemaking: An Analysis of Constitutional Issues that May Be Raised by Executive Order 12291,* 97th Cong. 1st sess., 1981.

165. *Reyes* v. *Wyeth Laboratories,* 498 F.2d 1264, 1274 (1974).

166. See Leventhal, "Environmental Decisionmaking," p. 550; O'Brien, "Seduction of the Judiciary;" Boyer, "Alternatives to Administrative Trial-Type Hearings"; Martin, "Procedures for Decision Making;" Arthur Kantrowitz, "Controlling Technology Democratically," *American Scientist* 63 (1975):505; Kantrowitz, "The Science Court Experiment," *Trial* (1977):44; Task Force of the Presidential Advisory Group on Anticipated Advances in Science and Technology, "The Science Court Experiment: An Interim Report," *Science* 193 (1976):653; and Howard Markey, "A Forum for Technocracy?" *Judicature* 60 (1977):365.

Part II
Policy Disputes

Biomedical Technology: The rDNA Controversy

Bonita A. Wlodkowski

Scientific Inquiry and Political Pluralism

In the fall of 1981, the most dramatic science-policy debate of the 1970s came to a close. The debate concerned the conduct of recombinant DNA (rDNA) research and focused on the question: Who should regulate that research? Between 1973 and 1978 the scientific community, state and local governments, and Congress vigorously debated the issue. Amid the debates in 1976, the National Institutes of Health (NIH) published guidelines regulating federally funded research. Initially controversial because of the risks of the research, rDNA research was also hailed as one of the most significant advances in twentieth-century biology and as the greatest medical innovation since antibiotics. The problems of assessing the technology, of weighing its risks and benefits, made resolution of the issue of regulation difficult and controversial. As the promise of benefits gradually supplanted the fear of hazard, and as some 200 universities engaged in research and commercialization expanded, the NIH progressively relaxed its restrictions.[1] In 1981, five years after it had promulgated guidelines, an NIH committee recommended the elimination of the restrictions. The NIH thereby concluded an unprecedented controversy in regulatory politics and science policy.

This chapter traces the history of the controversy. In chronologizing the major events, attention is given to the diverse political arenas that the controversy transversed: the scientific community, state and local governments, Congress, and the bureaucracy. I do not suggest that only one arena was exclusively engaged in the debate at any given time or that each arena entered the debate in a strictly linear, sequential fashion, but that is roughly how the debate proceeded. More important, my examination draws attention to when the issues reached various political agendas and focuses on how political institutions responded, as well as illuminates the pluralistic and polycentric nature of the rDNA controversy.

Following the historical account of the controversy, the chapter exam-

The author gratefully acknowledges the research support provided by the Council for Research in the Social Sciences and expresses appreciation to Patrick J. Hennigan, of Columbia University, for helpful comments on an earlier draft.

119

ines the political issues that gained prominence during the debate. The wide-ranging ramifications of the rDNA technique and its uncertain risks and benefits raised fundamental issues of science and technology policy. Fore-most among the issues were the role of public participation and the nature of governmental regulation of scientific inquiry. Underlying all public poli-cies are competing social and political values, but to an extraordinary degree, the rDNA controversy intensified and laid bare competing societal and political values: How should the values of scientific expertise and freedom be accommodated with the democratic ideal of self-governance?

This basic political issue invites polemics and was not conclusively resolved in the rDNA controversy. Indeed, as technological developments produced new information about risks and benefits, this issue became less prominent and critical in the rDNA debate. The potential societal effects of rDNA research are in the 1980s actually greater and more imminent than in the 1970s due to the increasing commercialization of rDNA technology. Though the controversy over rDNA research now appears settled, the pros-pect remains that government will again confront the issue of regulation as the commercialization and patenting of rDNA products expands.

The Scientific Community and the Emergence of Controversy

The scientific community first drew attention to rDNA research because of its great potential. Such research provides a powerful means of manipulat-ing the genetic material of living organisms.[2] Also known as gene splicing, it is the ability to transfer segments of genetic material to another organism. Restriction enzymes cut strands of DNA at precise points, and the DNA segment is recombined, or attached, to the DNA of a vector, either a virus or a plasmid. The recombinant molecule, consisting of both the DNA material and the vector, is then introduced into a host cell, usually a bac-terium. The host cell that was first used was *Escherichia coli* (*E. coli*), a natural inhabitant of the human intestine.

Clones of identical cells can be formed with each containing copies of the original rDNA molecule. Furthermore, genes inserted by the rDNA technique may be expressed—that is, a cell's biological machinery will syn-thesize the protein whose structure is specified by the gene; in this way protein products may be formed.

The implications of rDNA research embrace both benefits and risks. The possibility exists of installing new genetic information to compensate for genetic deficiencies and to create new forms of life. Another benefit lies in the opportunity to manufacture many biochemical substances more effi-ciently. Creating a new organism, however, may unintentionally produce disease, increase resistance to antibiotics, or disrupt the evolutionary process and thus pose dangers. The risks of rDNA research largely depend on the kind of genetic material introduced and the survivability of the cell outside the laboratory. This fear of risk caused the controversy within the scientific community.

The rDNA controversy publicly surfaced at the Gordon Conference on Nucleic Acids in June 1973, but its origins date two years earlier. An experiment in Paul Berg's laboratory at Stanford University aroused the concern of another researcher, Robert Pollack, during the summer of 1971. Berg's prototype gene-splicing experiment involved a monkey tumor virus, which is in some ways similar to human cancer, and *E. coli*. Fearful that the combination might escape from the laboratory and cause disease, Pollack prevailed upon Berg to postpone the experiment. Though initially skeptical, Berg eventually postponed the experiment because he could not deny the possibility of any risk. Berg and Pollack subsequently organized a conference on the safety of experimenting with tumor viruses in January 1973. This conference, now sometimes referred to as Asilomar I, revealed the limited state of scientific knowledge about biohazards. A second meeting was canceled due to the lack of information on biohazard risks.[3] Five months later, however, new findings reported at the Gordon Conference emphasized the urgent need for further study of rDNA research and its attendant risks.

At the Gordon Conference, Herbert Boyer described the gene-splicing technique that he and others had refined and its successful use in transferring the gene for penicillin resistance to *E. coli*.[4] Several participants requested that the conference chairpersons (Maxine Singer of the NIH and Dieter Söll of Yale) allocate time for discussion of the safety issues pertinent to the research. In the fifteen minutes that were formally arranged for discussion, it was decided to send a letter to the National Academy of Sciences (NAS) and the National Institute of Medicine (NIM), requesting that they study the risks of rDNA research. The conferees also decided, by a narrow vote of 48 to 42, for publication of the same letter in *Science*.[5]

Meanwhile, gene-splicing techniques became widely accessible to other scientists, and the plasmid used as a vehicle for carrying the DNA was made available to researchers. Stanley Cohen first used the plasmid, and he and his associates required that other researchers take precaution to minimize biohazard risks. Specifically, researchers could not share the plasmid with other laboratories or introduce it into tumor viruses or produce combinations that would render *E. coli* resistant to antibiotics without first informing Cohen. A system of self-regulation thus emerged but did not succeed in confining the plasmid to certain laboratories. Better controls appeared to be needed.[6]

The Gordon Conference letter, written by Maxine Singer and Dieter

Söll, addressed the problems of controlling biohazards. The publication of the Singer-Söll letter in *Science* brought the issue before the public. Noting that prudence, not the existence of known hazards, had prompted the letter, the authors requested that the NAS and NIM appoint a committee to study potential biohazards. The NAS responded by asking Paul Berg to chair a study committee. In July 1974, the Berg committee made recommendations that were also published in *Science*.[7] Concerned that artificial rDNA molecules could prove biologically hazardous, the committee called for a voluntary deferral of two specific types of experiments and recommended that others should be carefully considered before being undertaken. The committee also requested that the director of the NIH establish an advisory committee to evaluate the hazards, develop procedures to minimize risk, and devise guidelines for their implementation. Finally, the committee stated its intention to convene an international meeting that would review the state of rDNA research and discuss appropriate action.

The recommendations were limited—indeed, the scientists intended a pause, not a full moratorium—but the letter constituted a dramatic and significant event in the unfolding controversy. Though the issues had been publicized during the Gordon Conference and by the Singer-Söll letter, Berg's letter, calling for voluntary restraints and inviting governmental intervention, created an uproar within the scientific community. Nonetheless, researchers reportedly adhered to the recommendations of the Berg letter. In the United Kingdom, for example, possibly as a result of recent laboratory-caused smallpox deaths, the moratorium was strongly endorsed.[8] In the United States, on 7 October 1974, the NIH established the DNA Molecule Program Advisory Committee to "investigate the current state of knowledge and technology regarding DNA recombinants and recommend guidelines on the basis of the research results."[9]

The second Asilomar conference, as recommended in the Berg letter, was held in February 1975 and was attended by 155 scientists from around the world. Although the ethical implications of the research had been of growing concern, Asilomar's organizers adroitly identified its mandate as technical, not ethical or political. The Asilomar conferees decided that there was a "significant but not yet definable and measurable potential risk in certain types of recombinant DNA experiments."[10] The proposed safeguards relied on two principles of containment. The first, biological containment, referred to the use of bacterial hosts unable to survive in the natural environment and vectors capable of growing only in specified hosts. The second, a physical containment, was "exemplified by the use of suitable hoods or where applicable, limited access or negative pressure laboratories."[11] Accordingly, the scientists agreed on a strategy that matched levels of containment with types of rDNA experiments and that also deferred experiments presenting more-serious biohazard risks.

The general principles adopted at Asilomar reflected a moderate, cautious consensus. (Neither serious debate on discontinuing the research nor proposals for extensive governmental regulation occurred. The scientists who convened at Asilomar clearly viewed the moratorium as temporary but were of two different predispositions. One group expressed concerns about the research and believed that the scientific community was compelled to take appropriate action to minimize risk. A second group opposed any restrictive action but feared that silence from the scientific community would encourage governmental intervention. The first group prevailed at Asilomar. It did so because no one could disclaim any possibility of risk and because most participants realized the political value of achieving a consensus within the scientific community. For some, the value of the consensus was purely symbolic. For others, Asilomar was more. Scientists who anticipated governmental action viewed Asilomar as a means of influencing the political agenda. In retrospect, Asilomar accomplished both objectives. It registered the ability of the scientific community to organize itself and to reach consensus on self-regulation. Scientists thus gained the respect and trust of the public. In addition, they defined the issue as a technical one, thereby assuring themselves the major role in future science-policy disputes. Indeed, Asilomar laid the groundwork for the development of the NIH guidelines.

Initial Regulatory Responses: The NIH (More or Less) Intervenes

The Asilomar consensus dissipated during the sixteen-month period of NIH formulation of guidelines for rDNA research. The guidelines were finally issued by the NIH Advisory Committee on 23 June 1976, but only after a prolonged and intense debate within the professional community and regulatory arena. There were three major drafts of NIH guidelines. The NIH committee, established in October 1974, held its first meeting in February 1975. During that meeting the committee recommended that the NIH adopt the Asilomar guidelines until it published its own. In May, at the second meeting, a subcommittee was appointed to develop new NIH guidelines. The subcommittee report, the Hogness Paper, named after the subcommittee chairman, David Hogness, triggered debate over the nature and extent of containment necessary for the research. A second version was drafted at the third meeting in Woods Hole, Massachusetts. Weaker than the guidelines recommended at Asilomar, the Woods Hole version prompted serious objections. A group of forty-eight scientists sent a strongly worded petition opposing the draft for its lower safety standards and attacking the narrow composition of the committee. The Genetics and Society Group of

Science for the People, a Cambridge, Massachusetts–based group, also opposed the draft. They attacked the composition of the committee and specifically the dominance of geneticists, which they said was comparable to "having the chairman of General Motors write specifications for safety belts."[12]

As a result of the debate, DeWitt Stettin, the head of the NIH committee, circulated the guidelines outside the committee and appointed Elizabeth Kutter, a scientist but not a geneticist, to prepare yet another draft of the guidelines. The Kutter draft was compared side by side with the Hogness Paper and Woods Hole version at a fourth meeting in La Jolla, California. The primary point of contention centered on the classification of rDNA experiments according to specific levels of biological and physical containment. Many researchers feared that their research interests would be jeopardized by requirements to alter the physical construction of laboratories that were imposed by higher levels of containment. Agreement was nonetheless reached at La Jolla and the committee produced the "NIH Proposed Guidelines for Research Involving Recombinant DNA Molecules," guidelines that were more detailed and slightly stronger than those of Asilomar.

By the time the guidelines were finally developed, the NIH had to confront growing dissention within the scientific community. Molecular biologists had become restive. Eager to proceed with unrestricted research, unaccustomed to public scrutiny, and convinced that the risks had been exaggerated, they regretted having raised their concerns about rDNA in the first place. Moreover, the committee was under increasing criticism that its membership was too narrow. Largely in response to this, the NIH director, Donald S. Frederickson, called a public hearing in February 1976 and also broadened the membership of the committee. The only substantial modification made in the guidelines, however, was inclusion of recommendations made by a safety expert from the National Cancer Institute regarding physical-containment procedures. While finalizing and before issuing the guidelines, the NIH established the Office of Recombinant DNA Activities and met with industry and other federal agencies. At that time, only seven companies were engaged in rDNA research and, though hesitant to commit themselves, they endorsed the guidelines. In September 1976, the NIH also published a draft environmental-impact statement in order to promote greater understanding of the guidelines.[13]

The NIH guidelines were designed to establish uniform standards and procedures for the conduct of rDNA research. Recognizing that the task was to "allow the promise of the methodology to be realized while advocating the considerable caution that is demanded by what we and others view as potential hazards," the guidelines built on the principles and strategy of containment agreed on at Asilomar.[14] First, physical containment was specified at four levels: P1, requiring standard microbiological practices;

P2, requiring standard microbiological practices plus the use of certain types of containment equipment and specific sanitation procedures; P3, requiring more-stringent microbiological practices, exit and entry procedures, and facility safeguards; and finally, P4, which surpassed the requirements of P3 by requiring special cabinets (using negative pressure), air filtration, decontamination tanks, and secondary safeguards such as shower rooms for personnel. Second, biological-containment levels designated three strains of *E. coli* according to their survivability outside the laboratory. Specific classes of experiments were discussed in terms of the levels, and some experiments, considered hazardous, were prohibited. Third, certain experiments were strictly forbidden. Experiments using dangerous toxins and recombinants in excess of ten liters were prohibited. The latter provision was designed to limit the possibility of accident by keeping quantities small, but it posed special problems for large-scale industrial development.

Responsibility for implementation of the guidelines was fragmented among the individual researcher, institutional biohazards committees (IBCs), and the NIH. IBCs were to advise on policies and to review and approve the facilities, procedures, and practices of research projects to the NIH. At the NIH, study sections and initial review groups reviewed applications for rDNA research projects and considered both scientific merit and biohazards risks. An rDNA advisory committee (RAC) was charged with advising the NIH director on technical problems, and NIH staff were assigned to coordinate and monitor the overall program.

The NIH became involved in the rDNA controversy at the request of the scientific community and produced guidelines consistent with the consensus reached at Asilomar. Success of the guidelines primarily rested on the voluntary compliance of individual researchers. The NIH provided little indication that it would police individual researchers, though that possibility was not precluded. As Clifford Grobstein surmised: "The effort of the Asilomar Conference to achieve primarily self-regulation is preserved in the guidelines, but the involvement of NIH places self-regulation within a framework into which grades of external regulation can be introduced."[15]

Public Participation in Science-Policy Disputes: Local-Community Responses

While the NIH wrestled with formulation of guidelines that were of national applicability, the issue of public control over rDNA research prompted lively debates in several local communities in the mid-1970s. Localities debated the conduct of rDNA research within their jurisdictions. The first major debate occurred in Ann Arbor, Michigan, in early 1975, but by 1977 several localities and states were studying the issue. Scientists were especially apprehensive about local involvement; yet their concerns about extensive local-government intervention proved to be ill founded. By late 1979, only two states (Maryland and New York) and four localities (Cambridge, Amherst, Berkeley, and Princeton) had taken any kind of action, and four states (California, Wisconsin, New Jersey, and Illinois) had deferred action. The localities were quick to study the issue but ultimately restrained in their actions. Local governments supported NIH guidelines and in no way impeded research. For example, the town of Amherst, Massachusetts, passed an ordinance requiring that rDNA research be conducted according to NIH guidelines. In New York, guidelines that largely conformed with those of the NIH were promulgated, and the health commissioner was empowered to regulate rDNA research through a certification process. In Cambridge, Massachusetts, some modifications were made to the NIH guidelines, but the additional requirements were not found to be intolerable by scientists. More telling, despite local attention and often intensive scrutiny, little attempt was made to enforce local ordinances.[16]

The local governments had little real effect on the conduct of rDNA research. They nevertheless made an important contribution to the DNA debate. Local debates marked the beginning of a "new socio-scientific awareness."[17] In particular, the University of Michigan community was the site of the most extended and thorough discussion of the social, philosophical, and ethical aspects of rDNA research. Not surprisingly, debate focused on issues of importance to the academic community—scientific inquiry, administration of research, and the rights and responsibilities of scientists in the conduct of research. Controversy emerged when the University of Michigan decided to build a P3 facility and convened committees to study the plan. The committees eventually supported the university's construction plan but only after a thorough review of the issues. Unlike the Asilomar conference, which confined itself to technical issues, the University of Michigan debate embraced all aspects of the rDNA-research problem and, in particular, the ethical issues attending the possibility of creating new forms of life. As it turned out, the technical capability of minimizing biohazards was not disputed, and the philosophical aspects of rDNA research proved the most controversial.

While the most intensive debate occurred at the University of Michigan, perhaps the most spirited and publicized discussion occurred in Cambridge, Massachusetts. The basic difference between the two debates was that in Ann Arbor, the controversy was confined to the university, the dominating force in that community. In Cambridge, the city council debated the issue. Scientists at Harvard University were polarized over the plans of one department to construct a P3 facility in the biological-laboratory building. Opponents to the plan feared that ants that had infested the building would act as vectors in spreading hazardous material. Mayor Alfred Vellucci learned of the internal controversy from a local newspaper and from two university scientists who visited his office. Particularly persuaded by the scientists, he called public hearings on 23 June 1976 (which coincidentally was the data the NIH guidelines were finalized). Vellucci's plan for a two-year ban on rDNA research was eventually opposed by the city council. They instead voted for a good-faith moratorium on more-dangerous (P3 and P4) types of rDNA experiments while a group composed of laypersons and scientists studied the matter. Six months later, the Cambridge Laboratory Experimentation Board lifted the moratorium and, with some additional conditions, endorsed the NIH guidelines, noting that "a predominantly lay citizen group can face a technical scientific matter of general and deep public concern, educate itself appropriately to the task and reach a firm decision."[18]

The events in Cambridge are important for what they reveal about local-governmental intervention in science-policy disputes. Mayor Vellucci and the City of Cambridge became involved only because of internal controversy at Harvard University and at the urging of two scientists. It is unlikely that the research issue would have surfaced for public comment in the absence of internal dissension. The deliberations of the city council permitted expression of local fears and concerns about public-health risks. Moreover, the events at Cambridge demonstrated both the desire of the public to study the issue and their ability to participate in debates of scientific and ethical complexity. Arguments by members of the scientific community against the public participation in science-policy disputes were undercut by the relatively sophisticated nature of local debates. Still, fears that localities would follow Cambridge's example and establish citizen-review boards that might produce differing local regulations across the country remained a source of concern for the scientific community. Indeed, it was precisely the fear of local-government regulation that prompted scientific associations to lobby for federal legislation preempting local controls.

Professional Coalitions and Congress: Up the Hill and Down

Sparked by the well-publicized local controversies, various scientific interest groups became actively involved in the rDNA-research debate during 1977. The Association for the Advancement of Science devoted sessions at its annual meeting to the subject. However, the forum convened by the NAS demonstrated the commitment of various groups. Some organizations—Friends of the Earth, the Environmental Defense Fund, the Natural Resources Defense Council, the Cambridge-based Science for the People, and the People's Business Commission—called for a moratorium on rDNA research. The forum more than recapitulated arguments heard at local levels. New issues such as the political and economic nature of worker safety and decision making gained dominance. The forum was the culmination of public concern with rDNA, and it reinforced the need for national action.

During 1976–1977, the rDNA debate shifted to Congress. Motivated by events at the local level and by the failure of the NIH and professional communities completely to allay fears and resolve issues, Congress began considering regulatory legislation. This was not, however, the first time that Congress had been alerted to the issue. In 1971, James Watson, whose discovery with Francis Crick of the structure of DNA in 1953 had laid the groundwork for the gene-splicing technique, spoke before a congressional committee. He drew attention to the importance of the research and its fairly immediate benefits. Congressional interest in the rDNA research was again sparked in April 1975 by Senator Edward Kennedy's hearings on the relationship between science and society.[19] Still, not until the issuance of the NIH guidelines did congressional action become a real possibility.

In July 1976, less than a month after finalization of the NIH guidelines, Senators Edward Kennedy and Jacob Javits expressed their concern to President Gerald Ford that the guidelines, in applying only to federally funded research, were too narrow:

> We were gravely concerned that these relatively stringent guidelines may not be implemented in all sectors of the domestic and international communities and that the public will therefore be subjected to undue risks. . . . [G]uidelines are necessary for privately funded research and application as well.[20]

They requested that President Ford issue an executive order compelling private-sector compliance with NIH guidelines and prodded the president to submit legislation to Congress if he thought it necessary. The issue of regulating private-sector-sponsored rDNA research was pursued at the second set of hearings called by Senator Kennedy in September 1977. Arguing that the rDNA issue was of a prototype of the problems society would face with technological developments, Senator Kennedy insisted that the real problem was not safety but the social consequences of scientific advances and technological breakthroughs.[21]

Shortly after Senator Kennedy's hearings, the secretary of the DHEW, Joseph Califano, established the Federal Interagency Committee on rDNA research. Headed by Donald S. Frederickson, the NIH director, and composed of representatives of various federal agencies involved in or affected by rDNA research, the committee considered extension of NIH guidelines to all publicly and privately funded research. Though federal agencies were complying with the guidelines, the committee recommended federal legislation in March 1977. Secretary Califano reiterated the necessity of uniform standards, stating that "there is no reasonable alternative to regulation under law. Only continued research, proceeding under strict safeguards, will tell us whether these restrictions must continue in force or whether they can be relaxed at some time in the future."[22]

During 1977, sixteen bills were introduced and four committees held twenty-five hearings on markup sessions. The major bills were introduced by Congressman Paul Rogers and by Senator Kennedy. Congressman Rogers's bill was probably the most acceptable to the scientific community since it extended coverage of NIH guidelines to all federal agencies and provided for local variation only in special circumstances. The second major bill, Senator Kennedy's bill, was the most radical and proved to be unacceptable to the scientists. Not only did it permit local controls without federal preemption but it also recommended creation of a new, independent commission, thus removing regulatory responsibility from the NIH over research it funded.

Congressional debates during late 1977 centered on the effects of DNA legislation on other basic research and on the tension between public-health protection and scientific freedom as well as on the risks of rDNA research. Congressional interest was at a high point, yet no bill reached the floor for a vote. Legislation, which once had seemed inevitable, lost support for two reasons. First, several scientific professional groups intensely lobbied against regulatory controls, maintaining their adherence to the self-regulation strategy endorsed by the Asilomar conference and permitted under NIH guidelines. Scientists' lobbying efforts were well organized, intense, and sometimes unpleasant.[23] Harlyn O. Halvorson of Brandeis University organized scientists and more than two dozen science organizations in order to convince Congress that health hazards were exaggerated and that the NIH guidelines were adequate. The second factor that undermined passage of legislation was the growing evidence that health risks were fewer than originally feared. In April 1977, a researcher at the University of Alabama reported that use of certain weakened strains of *E. coli* posed no health risk. Then, in the following July, an NIH conference at Falmouth, Massachusetts, concluded that laboratory use of *E. coli* K-12 (a weakened strain) was safe and that the biohazards associated with rDNA research were overstated. New research also showed that natural recombinants occur and thereby suggested that rDNA research was less revolutionary and less dangerous than had been thought. Influenced by the new findings and persuaded by scientists' lobbying efforts, Congress gradually abandoned its efforts to pass regulatory controls on rDNA research. By late 1977 even Senator Kennedy withdrew support for his bill, though several committees in both houses continued to study the need for legislation in 1978.

In retrospect, Congress chided the bureaucracy for its failure to resolve the rDNA controversy, but it too found certain issues irreconcilable. The politics of regulation and the nature of the science-policy dispute were greatly complicated by new findings such as the fact that some recombinants occur naturally. This finding indicated that Congress would have to specify what was to be regulated—a subject on which no agreement could be reached. No agreement either could be reached on the sharing of responsibilities between federal and local levels for regulation. Mechanisms for enforcing congressional directives and the extent of acceptable surveillance of scientific research, particularly in industry where trade secrets had to be protected, posed additional obstacles. In sum, Congress appeared to be overwhelmed by the technical and political controversies attending the regulation of rDNA research. Congressional inaction in turn encouraged scientific self-regulation and shifted the debate back to the executive branch and, specifically, to the NIH. On 1 June 1978, Senators Kennedy, Javits, Nelson, Stevenson, Williams, and Schweiker wrote to Secretary Califano, urging that deficiencies in the guidelines be remedied by executive action based on existing statutory authority.

Back to the Bureaucracy and toward the Commercialization of rDNA

Several factors contributed to the changing role of the bureaucracy in the dispute over regulating rDNA. In the late 1970s, the NIH's risk-assessment program produced findings suggesting that the health risks of rDNA had been exaggerated. Just as the risk-assessment findings had influenced Congress, so too, the NIH became persuaded that the risks associated with rDNA research were less than initially predicted. The Falmouth conference and other workshops sponsored by the NIH's risk-assessment program produced a consensus on the relatively low risks of rDNA research.[24] There was also considerable pressure for increased citizen participation in rDNA regulation. DHEW Secretary Califano played an important role in this regard: He decided not to delegate all responsibilities to the NIH, and when the NIH RAC's charter was renewed in 1978, he reserved the right to appoint new members to the committee. In September 1978, Califano called a public hearing that was subsequently credited with expanding public participation in the IBCs of the NIH. Pressure had been mounting from environmental groups and various scientific organizations to expand membership in the IBCs. In the revised guidelines of 22 December 1978, 20 percent of IBC membership was to consist of public representatives, and public access to IBC records was increased.[25] These changes proved significant since the IBCs were also delegated major responsibility for assuring compliance with the guidelines. The RAC's membership was also increased and broadened to include experts from various scientific areas. Perhaps, as a consequence of these changes, in January and again in April 1980, the NIH modified its guidelines, specifying the procedures for voluntary compliance, especially industrial compliance, and requiring research laboratories to appoint a biological safety officer and to establish a worker-health surveillance program.[26]

Still, perhaps more important than expanding public participation in NIH committees, the phenomenal growth of industrial rDNA research had a decisive influence on the NIH. Only a handful of companies was involved in genetic-engineering research in 1978, but by 1981, fifty new firms (with a collective investment of $400 million) were started and just as many established companies joined the biotechnology gold rush.[27] As of April 1980, the RAC had approved two industrial applications. Less than a year later, sixteen more had been approved.[28] Interestingly, the RAC felt unqualified to assess large-scale fermentation practices and confined its activities to the setting of containment levels.

In response to these changing circumstances, the NIH repeatedly relaxed its guidelines.[29] By the end of 1980, research institutions were no longer required to register projects, and the NIH no longer reviewed IBC decisions on most rDNA experiments. Containment levels were lowered, and 80–85 percent of the rDNA experiments once covered by the 1976 guidelines had their restrictions removed. Though the experiments prohibited in 1976 remained so in 1980, the vast majority of experiments could now be conducted under standards of good microbiological practice.

The movement toward deregulation and commercialization of rDNA research received further support in 1980 from the Supreme Court's ruling in *Diamond* v. *Chakrabarty*.[30] The case involved an appeal of the denial of a patent application by a microbiologist for the invention of a bacterium capable of breaking down multiple components of crude oil. In a narrow five-to-four decision, the Supreme Court broadly interpreted the patent law and concluded that forms of life developed in the laboratory were not excluded from the statute and thus were patentable. Chief Justice Burger rejected the government's assertion that federal patent law did not include living organisms, dismissing the contention that a so-called parade of horribles would result from advancing genetic research and permitting the patenting of new forms of organic life. The majority, however, was careful to underscore that the political branches, not the courts, should ultimately resolve controversies over the risks and limitations of rDNA research.

Regulating the Biomedical Industry

The Supreme Court's ruling dramatizes the growth of commercial applications of rDNA research and technology.[31] The NIH, the leading federal actor since 1976, has also, perhaps less vividly but no less significantly, encouraged the commercialization of rDNA technology. On 9 September 1981, the NIH's RAC recommended the elimination of guidelines for rDNA research.[32] This preliminary proposal for deregulation culminates the continual relaxation of the 1976 guidelines. The decision indicates the NIH's position that the risks of rDNA research are sufficiently low and the benefits sufficiently high for federal regulation of research to cease. The decision reflects other changes in rDNA research over the last decade. For example, no longer is rDNA research university based. Industrial rDNA research was always outside the NIH's jurisdiction, and thus the expansion of private-sector rDNA technology signified that an increasingly high percentage of rDNA research would not be bound by the NIH, whatever its decisions. The NIH's role in the science-policy dispute over rDNA has progressively diminished.

The diminished role of the NIH and rapid industrial expansion have encouraged further local action. Since 1979, three local governments have imposed ordinances to govern all rDNA research conducted within their jurisdictions. Industry has viewed local action with some discomfort but has not found the local ordinances overly restrictive.[33]

Whatever federal regulation occurs in the future will undoubtedly be formulated by the FDA, OSHA, and the National Institute for Occupational Safety and Health (NIOSH). These agencies have begun studying their respective roles and responsibilities in regulating rDNA technology in view of the expanding interest in commercial uses of rDNA. The FDA has principal federal responsibility for regulating the safety of food, drugs, and cosmetics. New products resulting from rDNA research will thus undoubtedly require FDA approval prior to their marketing. In April 1981, the FDA reported that it had begun assessing methods for review of DNA applications and predicted that it would regulate on a product-by-product basis.[34] OSHA and NIOSH have likewise been examining their roles. The OSHA is presently monitoring rDNA-research operations in response to worker complaints. The NIOSH has conducted studies with the EPA and the Center for Disease Control in an effort to "help define the potential for health hazard to workers participating in commercial rDNA applications."[35]

The potential commercialization of rDNA research and technology thus bodes a shift in governmental attention to the marketing of safe consumer products and to assuring worker safety in employing rDNA technology. Further, commercial expansion has marked the trend away from regulation of basic scientific research into rDNA. The rDNA controversy has undergone a transformation since it emerged as a public-policy issue in the mid-1970s. The arenas of debate and the degree of public concern evolved with changing technological developments and arenas of political debate. The pluralism and fragmentation of the political debates and arenas reflect the inability of any single institution adequately to resolve science-

policy disputes. Most notably, the political salience of the rDNA controversy subsided as perceptions of risk diminished. From 1973 through 1978,
the controversy polarized professionals, politicians, scientific groups, and
the public. Yet today, the rDNA controversy is less dramatic, and the focus
has shifted from the potential risks of basic rDNA research to the benefits
and, to a lesser extent, to the risks of commercial exploitation of rDNA
technology.

Most of the basic issues generated by the controversy nonetheless
remain unresolved (and perhaps unresolvable): the precise role of public
participation in science-policy disputes, the moral and ethical dilemmas of
creating and patenting new forms of life, and the nature and extent of permissible governmental regulation of science and technology. The rDNA
controversy emerged as a major issue of public policy due to problems of
assessing uncertain risks and to basic questions about the governance of
science. Scientific expertise and freedom was pitted against demands for
political accountability and popular sovereignty. The nature and extent of
governmental intervention and the respective roles of federal, state, and
local governments were debated. Each of these political issues, however,
was modified by changing scientific and public perceptions of the risks and
benefits of rDNA research. Indeed, the inherent complexity and difficulty
of resolving science-policy disputes like that over rDNA arises because of
the lack of scientific certainty about risks and the attendant value conflicts.

Politics and Risk Assessment

Science policy is formulated under conditions of uncertainty. The problem
of dealing with uncertainty and of the assignment of utilities to risks and
benefits was especially acute in the rDNA controversy.[36] From a political
perspective, one of the most significant features of the rDNA controversy
was the changing assessments of the risks of rDNA. In the early and mid-
1970s, scientists encouraged political controversy by warning that existing
or newly created dangerous recombinants would escape from laboratories
and disseminate disease. Physical- and biological-containment procedures
developed at Asilomar and established by the NIH were accordingly designed to reduce the likelihood of the risks attendant rDNA research. In
addition to inadvertent release of microorganisms, there was also fear of
deliberate misuse of recombinants for biological warfare or terrorism.
Today, both of these risks appear remote. The risks arising from the release
of an organism presumed safe but that turns out to be harmful, however,
remains a source of controversy and indeed exemplifies the basic problems
of risk assessment. As an OTA report concluded: "The risk of harm refers
to the chance of harm actually occurring. In the present controversy it has

been difficult to distinguish the possible from the probable.''[37] The NIH's risk-assessment program provided information on the long-term consequences of rDNA, and largely as a result of the NIH's assessments, the risk of rDNA research appears to be considerably less than initially believed. The remaining controversy concerns the risks attendant the commercial exploitation of rDNA technology.

Risk assessments inescapably involve complex technical and scientific data. Not surprisingly, scientists largely dominated the rDNA controversy because of their command of technical information. Risk assessment itself, however, does not decide the acceptable levels of risk; determination of what constitutes an acceptable level of risk and the burden of proof remains a political decision, subject to societal judgments. Risk assessments inform political decisions, and ultimately, a political decision underlies regulatory responses to science-policy disputes. Fundamentally, the rDNA controversy raises the question of accountability and authority: Who in society—the public, scientists, or elected representatives—should make regulatory decisions and by what arrangements?

The Scientific Community, Public Participation, and Governmental Intervention

Scientists usually work in an environment free of public and political scrutiny.[38] The technological applications of scientific knowledge are usually the subject of public scrutiny but not the basic scientific research itself. In the rDNA controversy, however, science and technology appeared interdependent. The process of scientific inquiry as well as the technical applications of rDNA were publicly questioned. Indeed, it was the "publicness" of the rDNA controversy that made this science-policy dispute so unique. Scientists alerted the public to the controversy, yet they had no intention of relinquishing authority over their work. As public concern grew, at first registered at the local level and then in Congress, scientists emphasized new findings and determinations of low risk in order to discourage governmental regulation. They also argued that freedom of inquiry was in jeopardy. Most observers, however, agree that scientific freedom was never threatened or even at issue. Significantly, neither Congress nor local governments appeared to want to infringe on scientific freedom. When accused of doing so, they immediately withdrew. The principle that emerged was that scientific freedom should be limited only when there is a demonstrable public risk.

The basic political issue remains how to accommodate scientific freedom with public participation and governmental intervention in science-policy disputes. As James D. Carroll has observed, public involvement in science policy has become more important as scientific and technological developments have become binding forces in advanced industrial societies.[39] A kind of participatory technology appears to have emerged in the rDNA controversy. Public participation occurred at the local level and within the regulatory arena at the NIH. Membership on the NIH's RAC consistently expanded to include broader representation of interests. The import of public representation, however, remains debatable. At the local level, citizens studied the issues but deviated little, if at all, from established NIH policy. Likewise, at the NIH, nonscientists appeared to wield little actual power because of the dominance of scientists.[40] The experience at both local and federal levels suggests that scientists inevitably dominate science policy. At best, the concept of informed consent appears to operate when members of the public participate in the regulation of science-policy disputes.

Also prominent was the relationship between the scientific community and the regulatory agency. In retrospect, the NIH expanded public participation while it diminished its regulatory role. Indeed, although the NIH was the leading regulatory agency throughout the rDNA controversy, it resisted a regulatory role. It found such a role incompatible with its primary mission of promoting basic scientific research. Perhaps not surprisingly, the NIH cast rDNA disputes in terms of public participation and voluntary scientific compliance, thereby avoiding what it perceived as an unpalatable regulatory role and taking action opposed by its primary clientele.

Regulation of rDNA research amounted basically to self-regulation by the scientific community. The NIH was reluctant to regulate and especially to make decisions on novel types of experimentation. A request to undertake the first gene-therapy experiment, for example, was referred from one NIH committee to another, with each unwilling to decide the issue until the other did.[41] From all available evidence, scientists respected the NIH guidelines except for four known infractions. Notwithstanding these incidents, the history of the rDNA controversy reveals on the one hand the ability of scientists to regulate themselves and, on the other hand, the reluctance and difficulties of the NIH to monitor and enforce its regulations (in one instance, the NIH learned of an infraction from a published account in *Science*). The scientific community and its self-regulation, moreover, extended NIH guidelines to the private sector, compensating somewhat for the NIH's limited jurisdiction. Scientists at Stanford University and the University of California, for example, held a patent for a gene-splicing technique.[42] They in turn extended the license for the technique to private firms but also require that the firms comply with NIH guidelines. In effect, scientists extended the guidelines where the NIH could not. The development, extension, and infractions of NIH guidelines underscore what was obvious during the rDNA controversy—that is, both the conduct and restraint of rDNA research largely remains the province of the scientific community.

Conclusion

Within the history of the rDNA controversy, the movement toward regulatory legislation and then away from any regulation, even as embodied in the NIH guidelines for federally funded research, was determined by the continuing interplay of the quest for scientific certainty, intense value conflicts, and demands for political compromise. These competing values and forces, as well as the inherent tensions between scientific expertise and political accountability, underlie most science-policy disputes and promise again to figure prominently in future regulatory politics over the commercial exploitation of biomedical technology.

Notes

1. U.S. Congress, House, Committee on Science and Technology, Subcommittee on Science, Research and Technology, *Genetic Engineering, Human Genetics and Cell Biology Recombinant Molecule Research,* Supp. Rep. II, Committee Print, 94th Cong. 2d sess., 1976, p. 6, 233; and Patrick J. Hennigan, "Regulating Bio-Medical Technology: The Case of Recombinant DNA Research" (Cambridge: Harvard Business School, HBS Case Services, 1981).

2. U.S. Congress, House, Committee on Science and Technology, Subcommittee on Science, Research and Technology, *Genetic Engineering, Human Genetics and Cell Biology, Biotechnology,* Supp. Rep. III, Committee Print, 96th Cong., 2d sess., 1980, p. 9.

3. June Goodfield, *Playing God* (New York: Random House, 1977), p. 92; and Michael Rogers, *Biohazard* (New York: Alfred A. Knopf, 1977). p. 38.

4. Stanley N. Cohen, "The Manipulation of Genes," *Scientific American* (July 1975):25; Stanley N. Cohen, Annie C.Y. Chang, Herbert W. Boyer, and Robert B. Helling, "Construction of Biologically Functional Bacterial Plasmids *in Vitro,*" *Proceedings of the National Academy of Sciences* (November 1973), p. 70.

5. Letter from Maxine Singer and Dieter Söll to the NAS and NIM, reprinted in *Science* 181 (21 September 1973):114.

6. Clifford Grobstein, *A Double Image of the Double Helix* (San Francisco: W.H.Freeman, 1979), p. 18.

7. Letter from Paul Berg, et al., to the editor, printed in *Science* 185 (26 July 1974):303.

8. Rogers, *Biohazard,* pp. 45–48; and Goodfield, *Playing God,* p. 114.

9. *Federal Register* 39 (1974):306.

10. Grobstein, *Double Image of Double Helix,* p. 27.

11. Paul Berg, David Baltimore, Sydney Brenner, Richard O. Roblin, and Maxine F. Singer, "Summary Statement of the Asilomar Conference on Recombinant DNA Molecules," *Proceedings of the National Academy of Sciences* (June 1975), p. 72.

12. Judith P. Swazey, James R. Sorenson, and Cynthia B. Wong, "Risks and Benefits, Rights and Responsibilities: A History of the Recombinant DNA Research Controversy," *Southern California Law Review* 51 (1978):1039; and Goodfield, *Playing God,* p. 26.

13. *Federal Register* 41 (1976):38, 426.

14. *Federal Register* 41 (1976):27, 911.

15. Grobstein, *Double Image of Double Helix,* p. 42.

16. U.S. Congress, House, *Genetic Engineering,* p. 42.

17. Goodfield, *Playing God,* p. 186.

18. Report of the Cambridge Laboratory Experimentation Review Board, "Guidelines for the Use of Recombinant DNA Molecule Technology in the City of Cambridge," (Cambridge, Mass., January 1977).

19. U.S. Congress, Senate, Committee on Labor and Public Welfare, Subcommittee on Health, *Examination of the Relationship of a Free Society and its Scientific Community,* Hearings, 94th Cong., 1st sess., 22 April 1975.

20. Swazey, Sorenson, and Wong, "Risks and Benefits," p. 1064.

21. U.S. Congress, Senate, Committee on Labor and Public Welfare, Subcommittee on Health, *Examination of the NIH Guidelines Governing Recombinant DNA Research,* Joint Hearings, 94th Cong., 1st sess., 22 September 1976.

22. U.S. Congress, Senate, Committee on Human Resources, Subcommittee on Health and Scientific Research, Statement of Joseph F. Califano, Jr., Secretary of Health, Education, and Welfare, 94th Cong., 2nd sess., 6 April 1977.

23. Nicholas Wade, *The Ultimate Experiment* (New York: Walker and Company, 1979), pp. 172–174.

24. U.S. Congress, House, *Biotechnology,* pp. 24–27.

25. *Federal Register* 43 (1978):60, 126.

26. *Federal Register* 45 (1980):6, 743; and *Federal Register* 45 (1980): 24, 968.

27. "Biotechnology Becomes a Goldrush," *The Economist* (13 June 1981):81; and Jeffrey L. Fox, "Genetic Engineering Industry Emerges," *Chemical and Engineering News* (17 March 1980):30.

28. Nicholas Wade, "How to Keep Your Shirt—If You Put it in Genes," *Science* 212 (April 1981):26.

29. *Federal Register* 45 (1980):77, 372.

30. *Diamond* v. *Chakrabarty,* 65 L.Ed., 2d 144 (1980).

31. Anthony J. Parisi, "The Industry of Life: The Birth of the Gene Machine," *The New York Times,* 29 June 1980.

32. Marjorie Sun, "NIH Plan Relaxes Recombinant DNA Rules," *Science* 213 (25 September 1981):1482.

33. "Can Gene-Splicers Make Good Neighboors?" *Business Week,* 10 August 1981, p. 32.

34. Michael Goldberg and Henry I. Miller, "The Role of the Food and Drug Administration in the Regulation of the Products of Recombinant DNA Technology," *Recombinant DNA Technical Bulletin* (NIH) 4 (April 1981):15.

35. "NIOSH's Proposed Program for Addressing Potential Hazards Related to Commercial Recombinant DNA Applications (June 1980)," *Recombinant DNA Technical Bulletin* (NIH) 4 (July 1981):66.

36. Ruth Macklin, "On the Ethics of *Not* Doing Scientific Research," *The Hastings Center Report* (December 1977):11.

37. Office of Technology Assessment, *Impacts of Applied Genetics: Micro-Organisms, Plants, and Animals* (Washington, D.C.: Government Printing Office, 1981), p. 200.

38. Robert Neville, "Philosophic Perspective on Freedom of Inquiry," *Southern California Law Review* 51 (1978):1128.

39. James D. Carroll, "Participatory Democracy," in *Science, Technology and National Policy,* Thomas J. Kuehn and Alan L. Porter, eds. (Ithaca: Cornell University Press, 1981), p. 416.

40. Susan Wright, "The Recombinant DNA Advisory Committee," *Environment* 21 (April 1979):2. For a further discussion of the issue of public control, see Dorothy Nelkin, "Threats and Promises: Negotiating the Control of Research," *Daedalus* (1978):107.

41. U.S. Congress, House, *Biotechnology,* p. 22; Nicholas Wade, "Gene Therapy Caught in More Entanglements," *Science* 212 (1981):24; and Wade, *The Ultimate Experiment,* p. 163.

42. "Costing Clones," *The Economist,* (8 August 1981):273; and Patrick J. Hennigan, "Commercialization of Recombinant DNA Research: Government-Business-University Relations" (Cambridge: Harvard Business School, HBS Case Services, 1981).

Assessment of Technologies for Determining Cancer Risks from the Environment

Office of Technology Assessment

Public Perception about Cancer and the Legislative Response

The importance of cancer in U.S. policies about disease is illustrated by the attention focused on cancer research. The first institute of the U.S. Public Health Service to be devoted to a single disease was the National Cancer Institute (NCI), established in 1937. Initially a freestanding institute, it was incorporated into the NIH, which was organized in the 1940s. Thirty-four years after the NCI's establishment, a nearly successful effort was mounted in Congress, in 1971, to separate the NCI from the NIH and to establish a national cancer authority, which would have set cancer research further apart from other biomedical-research activities. While the National Cancer Act of 1971 was unsuccessful in establishing a new authority, it elevated the NCI to bureau status, a higher organizational level than all other institutes at the NIH until the National Heart, Lung, and Blood Institute was also made a bureau. The 1971 legislation established a three-person cancer panel, appointed by and responsible to the president; no other disease has been singled out in such a way. The attention bestowed on cancer research reflects the importance of the disease to the public. It is the number-two killer in the United States, the number-one killer among people younger than 55, and the most dreaded disease.[1]

In the 1960s and 1970s, public and congressional interest in cancer prevention was spurred by associations' being drawn between environmental exposures and cancer. Congressional testimony mentioned associations between the environment and cancer, and several laws were enacted to provide federal agencies with regulatory mechanisms to reduce exposures to carcinogens.

Public fear and dread of cancer is not likely to decrease, and despite the

This chapter is extracted from a report by the OTA, *Assessment of Technologies for Determining Cancer Risks from the Environment* (Washington, D.C.: U.S. Government Printing Office, 1981).

139

current antiregulatory mood, Americans still favor health and environmental regulations. A survey of 2,000 people, commissioned by Union
Carbide in 1979, found continued public support of government efforts "to
protect individual health and safety and the environment," Seventy percent
of those surveyed favored stronger measures to protect workers from cancer, and 65 percent favored stronger measures to protect consumers from
cancer.[2]

Concern for cancer is likely to provide impetus for continued efforts to
reduce its incidence and to improve its treatment. Efforts to improve treatment are seen as highly desirable and excite little controversy, but efforts to
reduce cancer incidence by regulatory intervention generate great passion
about whether the expected benefits from the regulations justify their costs.
Part of the controversy about regulatory intervention, whether it is worth it
or not, flows from those uncertainties, but controversy also stems from the
fact that the regulations bring two societal goals into conflict: The majority
of people wants protection from carcinogenic risks and, at the same time,
wants to reduce regulatory costs and burdens. Choosing between these two
goals or reaching compromises between them will remain an important
point of contention in policies about the control of cancer.

Several federal agencies administer regulatory programs for the control
of carcinogenic and other health risks to humans from chemical substances.
These programs differ in their objectives and regulatory authority. Some
were designed by Congress to deal with several different types of risks,
including carcinogenic risks, in the work place or in consumer products,
and others were designed to protect humans and the environment through
control of toxic substances in air, water, and food.

Regulatory decision making for control of cancer risks to humans is
guided by specific legal mandates and administrative procedures and
depends on technical determinations concerning the existence and magnitude of risk.

Cancer Mortality and Incidence

Cancer occupies center stage in U.S. concern about disease because of its
toll in lives, suffering, and dollars. It strikes one out of four Americans,
kills one out of five, and as the second-leading cause of death, following
heart disease, killed over 400,000 people in the United States in 1979.
According to estimates from the National Center for Health Statistics
(NCHS), cancer accounted for about 10 percent of the nation's total cost of
illness in 1977. These numbers are distressing, but the impacts of cancer
extend beyond the numbers of lives taken and dollars spent. The human
suffering it causes touches almost everyone.

Cancer is a collection of about 200 diseases grouped together because of their similar growth processes. Each cancer, regardless of the part of the body it affects, is believed to originate from a single transformed cell. A transformed cell is unresponsive to normal controls over growth, and its progeny may grow and multiply to produce a tumor. Studies in human populations and in laboratory animals have linked exposures to certain substances with cancer. This knowledge of cancer's origins has led to the conclusion that preventing interactions between cancer-causing substances and humans can reduce cancer's toll.

Nationwide mortality data are used to answer questions about the number of deaths caused by cancer in the United States. Without doubt, the number of Americans dying from cancer has increased during the last century. Paradoxically, a major part of this increase resulted from improvements in public health and medical care. In years past, infectious diseases killed large numbers of people in infancy and during childhood. Now that improved health care has softened the impact of those diseases, many more people live to old ages when cancer causes significant mortality.

Cancer deaths are not evenly distributed among all body sites; the lung, colon, and breast accounted for over 40 percent of the total (see table 8–1). Changes in cancer rates over time also vary by body site. For this reason, discussion of cancer rates at particular body sites is more revealing than discussion of overall trends that mask changes at individual sites. Moreover, because some cancer-causing substances act at specific sites, more information about opportunities for prevention is obtained from the analysis of particular sites.

To permit the examination of cancer rates over time, standardization, a statistical technique, is applied to make allowances for a changing population structure. Standardization allows the direct comparison of single, summary statistics—for example, the mortality rates from lung cancer for the entire population in 1950 and 1981. In this chapter, mortality rates are standardized to the age and racial structure of the 1970 U.S. Census, unless otherwise specified.

Age-specific rates are also used extensively for examining trends. These rates measure the proportion of people in defined age classes who have developed or died from cancer and thus are unaffected by changes in the age structure of the population. Of greatest importance in detecting and identifying carcinogens, changes over time in younger age groups often presage future, larger changes in that group of people as they enter older age groups.

In general, cancer mortality rates are higher among nonwhite men than among white men. Differences between nonwhite and white women are less pronounced. The observed greater fluctuations in rates from year to year for nonwhites is consistent with the conclusion that reporting of vital statistics is poorer for nonwhites than for whites.

Greatest concern is expressed about the increasing trends. The largest

Table 8-1
Mortality from Major Anatomical Cancer Sites in the United States, 1978, All Races

Anatomic Site	Number of Deaths			Percentage of Total		
	Male	Female	Total	Male	Female	Total
All malignant neoplasms	215,997	180,995	396,992	100	100	100
Lung, trachea, and bronchus	71,006	24,080	95,086	32.9	13.3	24
Colon	20,694	23,484	44,178	9.6	13	11.1
Breast	280	34,329	34,609	0.13	19	8.7
Prostate	21,674	—	21,674	10	—	5.5
Pancreas	11,010	9,767	20,777	5.1	5.4	5.2
Blood (leukemia)	8,683	6,708	15,391	4	3.7	3.9
Uterus	—	10,872	10,872	—	6	2.7
Ovary, fallopian tubes, and broad ligament	—	10,803	10,803	—	6	2.7
Bladder	6,771	3,078	9,849	3.1	1.7	2.5
Brain and other parts of nervous system	5,373	4,362	9,735	2.5	2.4	2.5
Rectum	5,002	4,089	9,091	2.3	2.3	2.3
Oral: Buccal cavity and pharynx	5,821	2,520	8,341	2.7	1.4	2.1
Kidney and other urinary tract organs	4,809	2,916	7,725	2.2	1.6	1.9
Esophagus	5,552	2,030	7,582	2.6	1.1	1.9
Skin	3,537	2,511	6,048	1.6	1.4	1.5
All other	45,785	39,446	85,231	21.2	21.8	21.5

Source: Office of Technology Assessment.

increases since 1950 are in respiratory cancers (mainly of the lung, larynx, pharynx, and trachea), which are largely ascribed to the effects of smoking. Male respiratory-cancer rates began to rise about twenty-five years earlier than female rates, which reflects the difference in time when the two sexes adopted smoking. Further evidence for the importance of smoking in lung cancer is the recent decrease in lung-cancer mortality among men younger than 50. The percentage of men who smoke is known to have decreased during the last twenty years, and studies have shown that smoking cessation reduces lung-cancer occurrence. Additionally, changes in cigarette composition are thought to contribute to a reduced risk of lung cancer. Decreases among men now over 50 are not expected because those populations include a large proportion of long-time smokers who remain at high risk.

Death rates from prostate and kidney cancers among men have risen somewhat, and mortality rates from malignant skin tumors (melanomas) have increased in white men and women. Mortality from breast cancer, the number-one cancer killer of women, has remained relatively constant. Overall mortality from nonrespiratory cancers, (that is, excluding most cancers normally associated with smoking) has decreased in women and remained constant in men during the last thirty years.

The more-satisfying trends are those that are decreasing. The most striking, among both men and women, has been the great decrease in stomach cancer since 1930. Although usually ascribed to changes in diet, the reasons for the decrease are not known with any certainty. A decrease in uterine cancer within the last few decades is attributed to higher living standards, better screening tests for early cancer, and an increase in hysterectomies, which reduces the number of women at risk.

Identification of Carcinogens

Carcinogens can be identified through epidemiology—the study of diseases and their determinants in human populations—and through various laboratory tests. Currently, eighteen chemicals and chemical processes are listed as human carcinogens, and an additional eighteen are listed as probable human carcinogens by the International Agency for Research on Cancer (IARC), a World Health Organization agency. IARC conclusions, based on reviews of the worldwide literature, are accepted as authoritative by government agencies and many other organizations.

In the United States, Congress has directed the National Toxicology Program (NTP) to produce an annual list of carcinogens. The first list, published in 1980, was composed of the substances identified as human carcinogens by the IARC. The next publication is to be considerably expanded and will include usage and exposure data and information on the regulatory status of over 100 chemicals whether considered to be carcinogens or regulated by the federal government because of carcinogenicity.

Cancer epidemiology established the associations between the thirty-six substances and human cancer listed by the IARC as well as the carcinogenicity of smoking, alcohol consumption, and radiation. However, epidemiology is limited as a technique for identifying carcinogens because cancers typically appear years or decades after exposure. If a carcinogen were identified twenty years after its widespread use began, many people might develop cancer from it even though its use is then immediately discontinued. Certainly, those people who were identified in the study as having had their cancer caused by the substance would have been irreparably harmed. Epidemiology is complicated because people are difficult to study; they move from place to place, change their type of work, change their habits, and it is hard to locate them in order to estimate their past exposures to suspect agents.

Laboratory tests, which do not depend on human illness and death to produce data, have been developed to identify carcinogens. Currently, the testing of suspect chemicals in laboratory animals, usually rats and mice, is the backbone of carcinogen identification. The suspect chemical is administered to the animals either in their food, water, or air or (less frequently) by force feeding, skin painting, or injection. As the animals die, or when the survivors are killed at the end of the exposure period (which is usually the lifespan of the animal), a pathologist examines them for tumors. The number of tumors in the exposed animals is then compared with the number in a group of control animals. The controls are treated exactly as the experimental groups except they are not exposed to the chemical under test. The finding of a significant excess of tumors in the exposed animals compared with the number found in controls in a well-designed, well-executed animal test for carcinogenicity leads to a conclusion that the chemical is a carcinogen in that species.

The IARC reviewed the literature concerning 362 substances that have been tested in animals and considers the data sufficient to conclude that 121 are carcinogens. For about 100 others, the evidence of carcinogenicity was limited, indicating that further information is desirable but that the available evidence produces a strong warning about carcinogenicity. Data were insufficient to make decisions about the carcinogenicity of the remaining substances. The IARC review program is active and updates its findings periodically.

The reliability of animal tests, bioassays, depends on their design and execution. The NCI published guidelines for bioassays in 1976. Bioassays now cost between $400,000 and $1 million and require up to five years to complete. Clearly such expensive tools should be used only to test highly suspect chemicals, and much effort is devoted to selecting chemicals for testing.

Molecular-structure analysis and examination of basic chemical and physical properties are used to make preliminary decisions about the likelihood of a chemical's being a carcinogen and whether or not to test it. For instance, greater suspicion is attached to chemicals that share common features with identified carcinogens. Unfortunately, not all members of a structural class behave similarly, which places limits on this approach. In making decisions about whether chemicals should be tested further, scientists consider other data including any available toxicological information. These preliminary decisions may be critical because, if a decision is made to test a substance, nothing more may be learned about its toxicity. The wrong decision might result in a carcinogen's entering the environment and being ignored until it causes disease in a large number of people.

The most exciting new developments in testing are the short-term tests, which cost from a few hundred to a few thousand dollars and require a few days to months to complete. Such tests have been under development for about fifteen years, and most depend on biologically measuring interactions between the suspect chemical and the genetic material, DNA. The best known test, the Ames test, measures mutagenicity (capacity to cause genetic changes) in bacteria. Other short-term tests use microorganisms, nonmammalian laboratory animals, and cultured human and animal cells. Some measure mutagenicity and some measure the capacity of a chemical to alter DNA metabolism or to transform a normal cell into a cell exhibiting abnormal growth characteristics.

Many chemicals that have already been identified as carcinogens or noncarcinogens in bioassays have also been assayed in short-term tests to measure congruence between the two types of tests. Results from these validation studies vary, but up to 90 percent of both carcinogens and noncarcinogens were correctly classified by short-term tests. These figures are sometimes questioned because they were derived from studies that excluded classes of chemicals known to be difficult to classify by the short-term tests being evaluated. However, the International Program for the Evaluation of Short-Term Tests for Carcinogenicity concluded that the Ames test, in combination with other tests, correctly identified about 80 percent of the tested carcinogens and noncarcinogens. That study purposefully included some chemicals known to be difficult to classify by short-term tests, and it further demonstrates the promise of these tests.

Short-term tests now play an important role in screening substances to aid in making decisions about whether or not to test them in animals. The role of short-term tests is expected to increase in the future as more such tests are developed and validated. However, the eventual replacement of animal tests by short-term tests is probably some time away.

One factor likely to retard replacement of animal tests by short-term tests is the poor quantitative agreement between the two kinds of tests. Qualitative agreement, as measured in validation studies, is good—that is, a mutagen is very likely to be a carcinogen—but poor quantitative agreement means that a powerful mutagen may be a weak carcinogen or the other way around. Additionally, because some evidence supports the idea that the potency of a carcinogen in animals is predictive of its potency in humans, the poor agreement about potency between animals and short-term tests may inhibit wider use of the latter tests.

Extrapolation from Test Results to Estimate Human-Cancer Incidence

Extrapolation techniques are used to estimate the probability of human cancer from studying derived data. Extrapolation can be divided into two parts. Biologic extrapolation involves the use of scaling factors to make adjustments between biologic effects in small, short-lived laboratory animals and in humans. Numeric extrapolation is used to estimate the probability of cancer at doses below those administered to animals in a test and to estimate cancer incidence at exposure levels other than those measured in epidemiologic studies.

Some extrapolation models assume a threshold dose—a nonzero dose below which exposures are safe and not associated with risk. Individual thresholds may exist because not all individuals exposed to similar levels of carcinogens develop cancer, but such differences in sensitivity may also be explained by differences in luck rather than in biology. However, a population threshold that would define a risk-free dose for a group of people composed of diverse individuals, if it exists, cannot now be demonstrated. Federal agencies do not accept the idea of thresholds in making decisions about carcinogenic risks.

Numeric-extrapolation models differ in the incidence of cancer that they predict from a given exposure. Extrapolation models that assume that incidence at low-exposure levels is directly proportional to dose usually estimate higher incidences. Such linear models are conservative in that, if they err, they overestimate the amount of disease to be expected. All governmental agencies that use extrapolation employ linear models for predicting cancer incidence. Other models project risks that decrease more rapidly than dose, and they are advanced as alternatives to the linear model. The choice of a model is important because, if an acceptable level of risk were decided on, almost any other model would allow higher exposures than linear models.

Opinions differ about whether and how extrapolation methods should be used in estimating the amount of human cancer that might be caused by exposure to a carcinogen. For example, some individuals object to any use of numeric extrapolation. For them, identification of a substance as a carcinogen is enough to justify efforts to reduce or to eliminate exposure. Other people see extrapolation as useful to separate more-risky from less-risky substances. The most extensive use of extrapolation is recommended by people who urge that extrapolation methods be used to estimate quantitatively the amount of human cancer likely to result from exposures. Such estimates are seen as necessary by those who wish to compare quantitatively the risks and benefits from carcinogens.

The disagreements among the groups who hold different opinions about the use of extrapolation are vocal and current. A particular problem in quantitative extrapolation arises from the fact that different extrapolation models produce estimates of cancer incidence that differ by factors of 1,000 or more at levels of human exposure. Given such uncertainty, some labor and environmental organizations and many individuals refuse to choose one model or another for estimating the impact of a carcinogen on humans and oppose the use of quantitative extrapolation. Fewer objections are raised against choosing a model to order carcinogens on their likelihood of causing cancer. Regardless of which particular model is chosen, it should produce approximately the same relative ranking as any other.

Proponents of quantitative extrapolation argue that careful attention to the available data aids in choosing the correct model and reduces chances for error. Arguments about the applicability of these techniques will continue, especially because efforts to apply cost-benefit analysis to making decisions about carcinogens will require quantitative estimates of cancer incidence.

No convincing data now exists to dictate which extrapolation model is best for estimating human-cancer incidence, whether from epidemiologic data or animal data, or even that one model will be consistently better than all others. However, one particular model for estimating human incidence from animal data (linear, no threshold extrapolation, and relating animal and humans on the basis of total lifetime exposure divided by body weight) has been reported to estimate human-cancer incidence within a factor of 10 to 100 when compared to incidence measured by epidemiologic studies. While this agreement is gratifyingly good, data exist to make these comparisons for fewer than twenty substances.

Statutory Mandates

Regulations of carcinogenic substances are designed to reduce health risks. The laws that require such regulations differ in whether they direct regulators to consider only health risks or both health risks and other factors. The other factors to be considered may include the costs of reducing the exposure and of foregone benefits from reduced availability of the substance.

Table 8-2 lists ten laws under which some action has been or may be taken to reduce exposure to carcinogens. Of the applicable statutes, the Federal Food, Drug, and Cosmetic Act (FDCA), the Clean Water Act of 1977, and the Toxic Substances Control Act (TSCA) specifically mention carcinogens. The remaining statutes provide for regulating all toxics, and carcinogens are included in that term. The list includes the laws most often discussed in relation to carcinogens but not all laws under which carcinogens might be regulated. For instance, laws governing transport of hazardous substances might be used to regulate carcinogens, and the DOA regulates carcinogens in poultry and meat.

Table 8-2
Public Laws Providing for the Regulation of Exposure to Carcinogens

Legislation (Agency)	Definition of Toxics or Hazards Used for Regulation of Carcinogens	Degree of Protection	Agents Regulated as Carcinogens (or Proposed for Regulation)	Basis of the Legislation	Remarks
Federal Food, Drug and Cosmetic Act (FDA)					
Food	Carcinogenicity for additive defined by Delaney clause	No risk permitted, ban of additive	21 food additives and colors	Risk	
	Contaminants	"Necessary for the protection of public health" Section 406 (346)	Three substances—aflatoxin, PCBs, nitrosamins	Balancing	
Drugs	Carcinogenicity defined as a risk	Risks and benefits of drug balanced.	Not determined	Balancing	
Cosmetics	"Substance injurious under conditions of use prescribed	Action taken on the basis that cosmetic is adulterated	Not determined	Risk; no health claims are allowed for cosmetics. If claims are made, cosmetic becomes a drug.	
Occupational Safety and Health Act (OSHA)	Not defined in act (but OSHA Generic Cancer Policy defines carcinogens on basis of animal-test results or epidemiology)	"Adequately assures to the extent feasible that no employees will suffer material impairment of health or functional capacity Section 6(b)(5)	20 substances	Technology (or balancing)	

					Basis of the Air-borne Carcinogen Policy
Clean Air Act (EPA)					
Stationary sources	"An air pollutant . . . which . . . may cause, or contribute to, an increase in mortality or an increase in serious irreversible, or incapacitating reversible, illness" Section 112(a)(1)	"An ample margin of safety to protect the public health" Section 112(b)(1)(B)	Asbestos, beryllium, mercury, vinyl chloride, benzene, radionuclides, and arsenic (an additional 24 substances are being considered)	Risk	
Vehicles	"Air pollutant from any . . . new motor vehicles . . . or engine, which . . . cause, or contribute to, air pollution which may reasonably be anticipated to endanger public health and welfare" Section 202A(a)(1)	"Standards which reflect the greatest degree of emission reduction achievable through . . . technology . . . available" Section 202(b)(3)(a)(1)	Diesel-particulates standard	Technology; Section 202(b)(4)(B) includes a risk-test for deciding between pollutant that might result from control attempts.	Section 202(b)(4)(A) specifies that no pollution control device, system, or element shall be allowed if it presents an unreasonable risk to health, welfare, or safety.
Fuel additives	Same as above [211(c)(1)]	Same as above [211(c)(2)(a)].	None	Balancing; technology based with consideration of costs, but health based in requirement that standards provide ample margin of safety.	A cost-benefit comparison of competing control technologies is required.
Clean Water Act (EPA)	Toxic pollutants listed in Committee Report 95–30 of House Committee on Public Works and Transportation. List from consent decree between EDF, NRDC, Citizens for Better Environment, and EPA (Section 307)	Defined by applying BAT economically achievable [Section 307 (a)(2)], but effluent levels are to "provide(s) an ample margin of safety" [Section 307 (a)(4)]	49 substances listed as carcinogens by CAG.	Technology	

Table 8-2 continued

Legislation (Agency)	Definition of Toxics or Hazards Used for Regulation of Carcinogens	Degree of Protection	Agents Regulated as Carcinogens (or Proposed for Regulation)	Basis of the Legislation	Remarks
Federal Insecticide, Fungicide, and Rodenticide Act and the Federal Environmental Pesticide Control Act (EPA)	One that results in "unreasonable adverse effects on the environment or will involve unreasonable hazard to the survival of a species declared endangered"	Not specified	14 rebuttable presumptions against registrations either initiated or completed; 9 pesticides voluntarily withdrawn from market.	Section 2(bb) Balancing "unreasonable adverse effects"	"Unreasonable adverse effects" means "unreasonable risk to man or the environment taking into account the economic, social, and environmental costs and benefits"
Resource Conservation and Recovery Act (EPA)	One that "may cause, or significantly contribute to, an increase in mortality or an increase in serious irreversible, or incapacitating reversible, illness; or pose a . . . hazard to human health or the environment . . ." Section 1004(5)(A)(B)	"That necessary to protect human health and the environment" Section 3002–04	74 substances proposed for listing as hazardous wastes	Risks; the administrator can order monitoring and set standards for sites.	
Safe Drinking Water Act (EPA)	"Contaminant(s) which . . . may have an adverse effect on the health of persons" Section 1401(1)(B)	"To the extent feasible . . . (taking costs into consideration)" Section 1412(a)(2)	Trinalomethanes, chemicals formed by reactions between chlorine used as disinfectant and organic chemicals. Two pesticides and 2 metals classified as carcinogens by CAG, but regulated because of other toxicities.	Balancing	

Toxic Substances Control Act (EPA)					
To require testing	Substances that "may present an unreasonable risk of injury to health or the environment" Section 4(a)(1)(A)(i)	Not specified	6 chemicals used to make plastics pliable	Balancing; "unreasonable risk"	
To regulate	Substances that "present(s) or will present an unreasonable risk of injury to health or the environment". Section 6(a)	"To protect adequately against such risk using the least burdensome requirement" Section 6(a)	PCBs regulated as directed by the law.	Balancing; "unreasonable risk"	
To commence civil action against imminent hazards	"Imminently hazardous chemical substance or mixture means a . . . substance or mixture which presents an imminent and unreasonable risk of serious or widespread injury to health or the environment." (Section 7)	Based on degree of protection in Section 6			
Federal Hazardous Substances Act (CPSC)	"Any substance (other than a radioactive substance) which has the capacity to produce personal injury or illness" 15 U.S.C. Section 2051	"Establish such reasonable variations or additional label requirements . . . necessary for the protection of public health and safety" 15 U.S.C. Section 2051		Risk	"Highly toxic" defined as capacity to cause death; thus toxicity may be limited to acute toxicity.

Table 8–2 continued

Legislation (Agency)	Definition of Toxics or Hazards Used for Regulation of Carcinogens	Degree of Protection	Agents Regulated as Carcinogens (or Proposed for Regulation)	Basis of the Legislation	Remarks
Consumer Product Safety Act (CPSC)	"Products which present unreasonable risks of injury. . . . in commerce," and "'risk of injury' means a risk of death, personal injury or serious or frequent injury," 15 U.S.C. Section 2051 "'Imminently hazardous consumer product' means consumer product which presents imminent and unreasonable risk of death, serious illness or severe personal injury," 15 U.S.C. Section 2061	"Standard shall be reasonably necessary to prevent or reduce an unreasonable risk of injury," 15 U.S.C. Section 2056	5 substances: asbestos benzene, benzidine (and benzidine-based dyes and pigment(s), vinyl chloride, TRIS	Balancing; "unreasonable"	Standards are to be expressed, wherever feasible, as performance requirements.

Source: Office of Technology Assessment.

Note: EDF = Environmental Defense Fund

NRDC = Natural Resources Defense Council

CAG = Carcinogenic Assessment Group

The earliest laws reflecting congressional concern about toxics centered on the food supply. Those laws, enacted around the turn of the century, established the FDA. In line with the importance society attaches to a safe food supply, the first law to apply directly to carcinogens was aimed at carcinogenic food additives. The Delaney clause, incorporated into the Food Additives Amendment of 1958, forbids the incorporation into food of any additive shown to induce cancer in humans or other animals.

The late 1960s and the 1970s saw the identification of carcinogens in various parts of the environment, and Congress provided legislative authority to regulatory agencies to reduce such exposures. The number and diversity of laws produces a balkanized federal regulatory effort. Whether or not carcinogen regulation would be better accomplished under fewer, broader laws is a question worthy of consideration but beyond the scope of the present assessment.

Bases for the Laws

Although many of the laws deal with other toxics in addition to carcinogens, the discussion here focuses on carcinogens to the exclusion of other health and environmental risks. The existence of the laws clearly states that Congress has seen cancer risks as deserving government attention. At the same time, despite the fact that some of the laws are attacked as proposing an unobtainable risk-free society, Congress has recognized that cancer-risk management can sometimes involve the balancing and comparing of risks against other societal goals. The ten laws in table 8–2 can be divided into risk-based laws (or zero-risk laws), which allow no balancing of health risks against other factors; balancing laws, which require balancing of risks against benefits of the substance; and technology-based laws, which direct regulatory agencies to impose specified levels of control.[3]

Risk-Based Laws

For this discussion, *risk-based* refers to legislation that provides for regulations to reduce risks to zero without considering other factors. The primary example is the Delaney clause, which specifies that carcinogenic food additives are to be eliminated from the food supply. Section 112 of the Clean Air Act and section 307(a)(4) of the Clean Water Act call for the reduction of exposures to levels that allow an "ample margin of safety." Because federal agencies do not accept threshold levels below which carcinogens pose no risk, strict interpretation of ample margin of safety would also require reductions to zero exposures. The Resource Conservation and Recovery Act (RCRA) is also risk based, but no regulation about carcinogens has yet been issued under it.

When it enacted the Delaney clause, Congress was aware that over 1,000 substances were present as additives in the U.S. food supply. Many of those substances had been poorly tested (if at all) for acute toxicity, and hardly any had been tested for chronic toxicity. Congress recognized that some of the additives might pose health problems, and the Delaney clause reflects its conclusion that no benefit could militate against banning a carcinogenic additive.

The most recent application of the Delaney clause was the proposed removal of saccharin from the "generally recognized as safe" list of food additives. This action, based on the finding that saccharin causes bladder cancer in rats, would then have resulted in the ban of saccharin from use as a food additive.[4] Few people question that saccharin is a rat carcinogen, and the Delaney clause is clear: Saccharin should be removed from the market of the basis of the animal study. Peter Hutt, former general counsel of the FDA, points out that a zero-risk approach such as the Delaney clause may work reasonably well so long as substitutes exist. When no substitute is available, controversy flares; it flamed in the case of saccharin.

Because the FDA had banned cyclamates on the basis of animal tests in 1969, the banning of saccharin would have meant that no nonnutritive sweetener would be on the market. No one could have the pleasure of sweetness without the cost of calories. Citizens who objected to the ban, aided by postage-paid postcards inserted into cartons of saccharin-sweetened soft drinks, deluged Congress with mail. Congress delayed the imposition on the saccharin ban and called for more studies. In 1980, it continued the delay.

At the request of Congress, both the OTA and the National Research Council (NRC) reviewed the scientific data and conclusions about saccharin and agreed with the FDA decision that saccharin is a carcinogen.[5] The congressional decision to forestall action on saccharin can be viewed as reopening the discussion about its earlier decision that benefits of carcinogenic food additives could not be weighed against their risks. It is aware of the evidence or risks (from animal studies) and benefits (from public outcry), and by delaying the ban, it is giving weight to the benefits.

Whether the saccharin moratorium presages an eventual voiding of the Delaney clause is an open question. Congress may retain the Delaney clause but, from time to time, exercise its prerogative to overrule agency decisions when it decides that benefits, whether measured or not, outweigh risks.

Comments are sometimes heard that the Delaney clause was appropriate for the state of knowledge in 1957 when it was enacted but that times have changed. Some observers say that, at the time, people usually agreed that few substances were carcinogenic and that those few could be eliminated from commerce with little difficulty. However, more and more substances, including many useful and some apparently essential ones, have been identified as carcinogens. Some people make a connection between these discoveries and the apparent turning away from risk-based laws such as the Delaney clause.

Despite the fact that some of the laws written in the 1970s were technology based or balancing, Section 112 of the Clean Air Act of 1970 and the RCRA of 1976 are risk based. They direct that risks to health be reduced or eliminated without specifically calling for consideration of other factors. An important consideration in these two laws is that they deal with pollutants. Pollutants benefit no one, and their reduction or elimination improves the environment and public health.

A problem arises because eliminating the pollutants costs money. Furthermore, in the case of carcinogens, the evidence that a substance poses a risk is not always accepted by everyone. As a result, the cost of reducing exposure to a pollutant is often offered as an argument against the projected health benefits expected from regulation, and suggestions are made that the costs be considered against the benefits before regulation, even in cases where the law does not call for such considerations.

Balancing Laws

The balancing laws, such as the Consumer Product Safety Act (CPSA), the Federal Insecticide, Fungicide and Rodenticide Act (FIFRA), and the TSCA put a qualifying word like *unreasonable* in front of the word *risk*. This construction implies that some risks are to be tolerated and, in practice, means that risks from a substance are to be weighed against other factors in the process of deciding whether and how to regulate. The TSCA requires that "the benefits . . . for various uses, . . . the economic consequences of the rule, . . . the effect on the national economy, small business, technological innovation, the environment, and public health" [§6(d)(D)] be considered in deciding whether a substance does or does not pose an unreasonable risk.

Balancing is equated with some kind of comparison of benefits and costs, but none of the laws explicitly requires formal benefit-cost analysis. For instance, the Committee on Interstate and Foreign Commerce report on the TSCA says that "a formal benefit-cost analysis under which a monetary value is assigned to the risks" is not required.[6] Also, a court decision about an action taken under the CPSA declared that the Consumer Product Safety Commission "does not have to conduct an 'elaborate cost-benefit analysis' to conclude that 'unreasonable risk' exists."[7]

All of the laws provide for the regulation of carcinogens that threaten human health. In the case of the balancing laws, Congress requires that other considerations be balanced against the health risk. In practice, health risk signals an agency that it should consider regulation; the stringency of that regulation is at least partially determined by balancing.

Technology-Based Laws

The Clean Water and Clean Air Acts are essentially technology based. For instance, the Clean Air Act directs the administrator of the EPA to reduce particulate emissions to some percentage of existing levels. These regulations may be technology forcing because new techniques may be required to achieve the reduction. In other cases, the laws specify that pollution control is to be achieved by using best practical technology or best available technology. Such regulations do not force new technology but bring all control efforts up to standards established by existing control technologies.

An important consideration of the technology-based laws is that the EPA has not yet been required to produce studies to show that the imposition of new standards will improve public health. Imposition of the standards reduces exposures, and in the case of carcinogens, given a nonthreshold approach to carcinogenic risks, it follows that reducing exposures should improve public health.

The Occupational Safety and Health Act requires:

> The secretary, in promulgating standards . . . shall set the standard which most adequately assures, to the extent feasible, on the basis of the best available evidence, that no employee will suffer material impairment of health or functional capacity. . . . In addition to the attainment of the highest degree of health and safety protection for the employee, other considerations shall be . . . the feasibility of the standards. [§6(b)(5)]

In the sense that *feasible* has a technological meaning, the Occupational Safety and Health Act can be considered as a technology-based law. However, the Supreme Court may issue a decision, in a case involving Occupational Safety and Health Act standard for exposure to cotton dust, that will determine whether or not benefits and costs have to be calculated to justify the standard. [The Supreme Court, on 17 June 1981, held, in *American Textile Manufacturers Institute, Inc.* v. *Donovan,* that Section 6(b)(5) does not require the OSHA to conduct a cost-benefit analysis.]

Freeman has argued that the technology-based laws require balancing, pointing out that best practical technology implies balancing.[8] How else can *practical* be defined? Likewise, deciding what is best available technology involves balancing costs of the technology against the expected gains.

David Doniger cites a significant balance between technology-based and balancing laws.[9] He suggests that once a hazard is identified under a technology-based law, the next step is to determine the best means to control it and then to decide if there are any compelling reasons to back off from the best means. Under a balancing law, he says, once a hazard is identified, the next step is to quantify the risks it presents in order to balance those against costs of control.

The tripartite division of the laws—risk, balancing, technology—while useful, does not neatly describe all the laws when subjected to closer inspection. Complex laws contain sections that have different bases, and carcinogen regulations are typically developed under risk-based or balancing sections of those laws.

An Example of Balancing

The FDA can balance costs and benefits in regulating carcinogens in food except when the carcinogen is a food additive. An example of that balancing is the FDA regulation of PCBs in fish.[10] The FDA considered three possible levels for PCBs in fish from the Great Lakes (see table 8–3). Fish that contain PCBs up to the FDA-established tolerance level can be sold; those having more PCBs cannot be sold.

Few (perhaps no) people dispute that PCBs are a human-health hazard. The acceptance of that fact is amply demonstrated by the TSCA [§6(e)] which directs that PCBs be regulated. The information in table 8–3 may illustrate that once a risk is accepted as real—that is, worthy of regulatory attention—the stringency of the regulation is set by economic or other factors. It is reasonable to assume that more cancer would result from PCBs at 5 ppm than at 1 ppm, but given uncertainties of quantitative risk assessment, it is difficult or impossible to accept that the projected number of cases is accurate. Nevertheless, the FDA decided that the balance between public-health protection and loss of food is properly struck by a 2 ppm tolerance.

Table 8–3
An Example of Balancing Cancer Risk versus Revenue Loss:
The FDA's Setting of a Tolerance for PCBs in Fish

Proposed Tolerance in ppms	Number of Projected Cancer Cases/Year	Estimated Loss of Revenue
5	46.8	$ 0.6 million
2	34.3	5.7 million
1	21	16 million

Source: Office of Technology Assessment.

The Feasibility and Limits of Benefit-Cost Analysis

Traditional benefit-cost analysis is an economic tool that requires listing all benefits and costs and assigning a dollar value to each. Carcinogen regulations involve the ultimate considerations—life and death—and opinions differ about whether or not a monetary value can be placed on life. In whatever way that controversy is settled, certain comments can be made about the usefulness of benefit-cost analysis.

In the case of a carcinogen regulation, expected benefits include the health gains from reducing exposure to the agent, and uncertainty is attached to the calculation of these gains. Presenting the expected health gain as a number does not add certainty to the estimate, but it adds to the estimate's importance. A frequent observation is that caveats and reservations attached to the numbers in the analysis are lost. The number of premature deaths to be averted (lives saved) as well as the number of dollars to be spent to achieve the benefit take on lives of their own and are encumbered by statistical reservations about accuracy. In this way, with no more information behind them, the numbers become more certain in the public's and the decision makers' minds. Too much reliance on imprecise numbers becomes a special problem in benefit-cost analysis that reduces the benefits and costs to dollar figures. This problem is commonly acknowledged, but a practical solution is not readily apparent.

Many criticisms are directed at a benefit-cost analysis. Lester Lave claims that a well-done analysis will always favor the status quo; change costs money.[11] Change can be produced by going either from regulation to no regulation or from no regulation to regulation. Lave goes on to say that it would be surprising if the present state is truly the pinnacle of social evolution. If such a tilt toward the status quo exists when all measurements and calculations are accurately done, it is easy to imagine how a bias on the part of the analyst could affect the analysis. Not all economists agree with Lave's statement that benefit-cost analysis will always favor the status quo, and Freeman cites the NAS 1974 study on air quality and automobile-emissions control as an example of cost-benefit analysis that favored tighter controls.[12]

Two other criticisms are directed at benefit-cost analysis. First, equity considerations are not a part of benefit-cost analysis. For instance, the health benefits from a regulation accrue to those individuals whose risks are decreased; the cost of reducing the risks is borne by those who pay for the control devices or procedure. However, the situation before the control is implemented has the exposed people bearing a health risk that spares anyone in society from having to pay to reduce the exposure. The technique of cost-benefit analysis is silent about whether either case is just.

Another difficulty encountered with benefit-cost analysis is its inability to consider intergenerational effects. As a specific example, many substances are both carcinogens and mutagens. Such a substance may cause cancer in people exposed to it and mutations in their germ generation. Quantifying genetic damage is at least as difficult as expressing the value of a life in dollars.

A quite different problem in dealing with the future is economic. The discount rate chosen to project monetary costs and benefits is very important to these analyses, but there is no agreement about the appropriate rate, especially in inflationary times.

A response to objections about valuing lives in dollar terms is seen in the suggestion of the Conservation Foundation that different types of analysis can be carried out under the benefit-cost rubric.[13] It suggests use of the term *single-value analysis* to describe benefit-cost analyses that express all items in dollar terms and *two-value analysis* for analyses that compare lives saved to dollars cost.

The Conservation Foundation found two-value analysis appropriate for carcinogens when the major concern is death. It is less easily applied when additional considerations like damage to an ecosystem are involved. *Multivalue analysis* is suggested as a method to consider three or more irreducible elements. It enables trade-offs to be made—for example, human-health risk versus costs of reducing exposure and ecosystem risk versus costs—but the analysis becomes more difficult. The claimed advantage of benefit-cost analysis (that it forces a detailing of what is being considered) is equally applicable to the one-, two-, and multivalue methods. The two- and multivalue methods involve balancing of health risk against other factors and fit within Lave's third framework of general balancing of risks and benefits.

While two-value or multivalue benefit-cost analysis may be attractive because of its rigor and its not placing a dollar value on life, it is not, strictly speaking, benefit-cost analysis. A benefit-cost analysis drives toward a single number, the quotient obtained when benefits are divided by costs. If the quotient exceeds 1, the benefits are greater that the costs, and the project should, on an economic basis, proceed; if it is less than 1, it should not. Two-value and multivalue benefit-cost analyses produce no such quotient. Instead they compare the benefits in one term (lives) versus costs (in dollars). The comparison is useful in a cost-effectiveness approach, where various approaches to a common goal can be ranked, but it does not produce a number that indicates yes or no.

A working panel of the NAS Committee on Principles of Decision Making for Regulating Chemicals in the Environment concluded that:

> The systematic application of the tools of decision analysis and benefit-cost analysis can provide the decision maker with a useful framework and language for describing and discussing trade-offs, noncommensurability, and uncertainty. This framework should help to clarify the existence of

alternatives, decision points, gaps in information, and value judgments concerning trade-offs.[14]

Decision analysis, as described in the NAS report, is a careful detailing of regulatory options, expected outcomes and uncertainties of risks, benefits, and costs. Its main contribution is to organize information for the decisionmaker to assist him in his unavoidable balancing task.

NAS committee endorsed the use of benefit-cost analysis and concluded that the technique is useful in making decisions but that it should not be the only consideration in the decision.[15] Furthermore, it did not recommend continued research to improve the techniques because "highly formalized methods of benefit-cost analysis can seldom be used for making decisions about regulting chemicals in the environment":

> If the market prices and shadow prices are fully utilized to value economic efficiency effects, the initial list of noncommensurate effects of the decision will have been reduced to:
>
>> Fully Commensurate economic efficiency benefits and costs measured in dollars; and
>
>> Noncommensurate effects, described and quantified in other units, of which the most significant are likely to be hazards to health and life, damages to the environment and ecosystems, and the distribution of benefits, costs, and hazards among individuals and groups.

The NAS definition of benefit-cost analysis does not require that all costs and benefits be expressed in a common unit (dollars), and it too would fit into Lave's third framework.

This assessment, which deals with the science and policy of making decisions about carcinogens, has emphasized uncertainties in determining risk. The Conservation Foundation and the NRC also draw attention to uncertainties in projecting costs of regulations.[16] Both sides of the benefit-cost analysis are difficult, subject to human error, and encumbered by uncertainties. Nevertheless, this method, whether one-, two-, or multivalue, has its advocates, and it is increasingly mentioned as at least a tool to be used by decision makers.

Regulatory Responsibility

Quantification of risks and benefits is never likely to be so precise that estimates of these variables can be plugged into a formula to produce a number that dictates a decision. Instead, after all the analyses, a decision about whether a risk is acceptable or unacceptable and reasonable or unreasonable will be made by a few individuals. Those judgments reflect societal values and would, ideally, be made by citizens as a whole. In our form of government, elected representatives are responsible for expressing societal values, and this responsibility is sometimes delegated by elected representatives to executive- or judicial-branch officials.

Executive-Branch Decisions

The size and complexities of government have resulted in elected representatives delegating authority to make judgments about acceptable risk to executive-branch agencies. A number of commentators, including Judge Howard Markey, point out that many executive-branch officials are civil servants with almost lifetime tenure. While high-level appointees like the EPA administrator are at some risk if they make poor decisions, tenured civil servants are not. Citizens who feel aggrieved as a result of executive-branch decisions have little opportunity for redress.[17]

Robert Field cites a number of legal scholars who contend that "vague mandates" such as the TSCA's directions to reduce or eliminate unreasonable risks give too much discretion to agencies:

> If regulatory decisions are to be broadly acceptable, the governing statutes must do more than provide for decisions about what is safe or what is an unreasonable risk. They must also do more than merely list factors to be considered. The legislative process must make the basic value judgments and tell the agencies how to make the necessary trade-offs. . . .
>
> Insofar as statutes do not effectively dictate agency actions, individual autonomy is vulnerable to the imposition of sanctions at the unruled will of executive officials, major questions of social and economic policy are determined by officials who are not formally accountable to the electorate, and both the checking and validating functions of the traditional model are impaired.[18]

Baram draws attention to the frequent absence of congressional direction about what factors agencies are to consider in reducing risk.[19] Absence of that direction, he says, causes the agencies to face extensive litigation.

Judiciary-Branch Decisions

Judge Howard Markey expresses concern about the judiciary's becoming too involved in making decisions about acceptable risk.[20] He sees such decisions as best made in the political arena by officials responsible to the electorate. Judges, he points out, often enjoy lifetime tenure and are more removed from the electoral process than executive-branch officials.

Vague definitions like unreasonable risk that occur in the balancing laws are seen as inviting legal challenge and judicial involvement in risk decisions. The courts have usually given great weight to "agency procedural safeguards, substantiality of evidence, and consistency," but two recent developments may alter such preference.[21] The Bumpers amendment would erase the judicial preference shown to agency expertise, and the courts would entertain challenges to agency expertise. Judicial involvement is also increased because of industry challenges to agency rules and because of consumer, environmental, and labor-organization challenges to agencies for not regulating. Whatever the source of challenge, judicial reviews have an impact on what level or risk will be acceptable. Should a court modify an agency-decided risk level, it is reasonable to assume that the agency, in making its next decision, will consider the courts' decree.

Legislative-Branch Directions

Congressional attention to details about what to balance and how to balance are seen as solutions to some of the problems of the regulatory agencies. Interestingly, regulatory-agency attorneys interviewed by Field favored balancing laws.[22] They see their agencies as well able to do an adequate job of balancing risks and benefits and evidently do not share the concerns about vague mandates.

Attorneys for four environmental interest groups were also interviewed.[23] Three were opposed to balancing laws and favored that the Congress impose clearer mandates for regulatory action. Vague balancing laws were seen as favoring industry because of its greater resources for influencing decisions in the agencies and in the courts. The fourth environmental attorney suggested that lack of money hampered public interest-group representation and that federal funding for preparation of their cases would ease their difficulties.

Clearly, different laws impose different standards, and the balancing laws are not specific about what is to be balanced. Litigation results from these characteristics of the laws. Congressional intervention to define standards and to detail what to balance should reduce litigation from those sources.

However, there is no guarantee that clearer directions from Congress will eliminate litigation—for example, the FDA's banning of the artificial sweetener, cyclamate, begun in 1959, was contested until 1980 (and may be reopened). Of all the carcinogen laws, the Delaney clause, which was the basis of the cyclamate ban, provides the clearest definition of a carcinogen, and it is the simplest in the sense that it allows no balancing. Nevertheless, administrative-law hearings contesting the quality of the evidence about carcinogenicity were held off and on for a decade. The FDA decision was upheld.

Congress cannot engage in the day-to-day business of agencies; it does not have the time. It must delegate authority to the agencies. If it shares the opinions of some observers that it does not provide sufficient direction to the agencies about acceptable risk and balancing, then it might provide the direction or it might more often exercise its right to intervene in regulatory activities.

Some risks are inherent in either remedy. On the one hand, overly strict directions, which provide many points for judicial review, might so encumber the agencies that no preventive regulatory action is possible. On the other hand, Congress cannot intervene too often in regulatory matters without hobbling its capacity to deal with other issues.

Regulated Carcinogens

Table 8–4 lists 102 substances and categories of substances regulated under the laws discussed in this chapter. In every case, some evidence existed to indicate that the substance is a carcinogen. In many cases, evidence about the substance was generated in the NCI bioassay program[24] and/or evaluated by the IARC.[25] The left-hand columns of the table describe what conclusions were drawn about the human and animal substances by the NCI and IARC.

The IARC has classified eighteen substances or processes as human carcinogens and another eighteen as probable human carcinogens. The data in table 8–4 show that twenty-one of those chemicals are regulated. Each of those twenty-one is identified by a C, carcinogen, or PC, probable carcinogen, in the H, human-evidence, column under the IARC.

About one-third of the chemicals tested by the NCI or reviewed by the IARC present sufficient evidence to conclude that they are carcinogens in animals and that they can therefore be assumed to present a carcinogenic risk for humans.[26] Chemicals from those classes are indicated with an S on the table in the A, animal-evidence, column; the I classification means the available evidence is limited but presents a strong warning of carcinogenicity.

About half of the chemicals reviewed by the IARC and/or tested by the NCI presented neither sufficient nor limited evidence of carcinogenicity. Only one chemical for which there is only inadequate (I) evidence of carcinogenicity appears in table 8–4. It is not possible to decide from the data in the table that risky chemicals are being regulated at the proper pace, but the data do lead to the conclusion that nonrisky chemicals (as judged by the IARC and NCI) are not often regulated. The conclusion then suggests that regulations are not so haphazardly drawn as to regulate large numbers of chemicals that present no or very little risk.

Table 8-4
Substances Regulated as Carcinogens under Various Acts

	Evaluation by				Statutes							
	NCI	IARC										
Chemical	A	H	A	CAA	CWA §307	CWA §311	SDWA	FIFRA	OSHA	FDCA	CPSA
2-acetylaminofluorene (2-AAF)				—	—	—	—	—	—	—	—
Acrylonitrile		PC	S	C	RR	L	—	V	R	R	—
Aflatoxin		PC	S	—	—	L	—	V	—	R	—
Aldrin	L		L	—	RR	—	—	V	R	—	—
4-aminobiphenyl		C	S	—	—	—	—	—	R	—	—
Aramite			S	—	—	—	—	V	—	—	—
Arsenic		C	I	P	RR	—	R[a]	R	R	—	—
Arsenic compounds		C	I	—	RR	L	—	R	—	—	—
Asbestos		C	S	R	RR	—	—	—	R[a]	—	R
Benz(a)anthracene			S	—	RR	—	—	—	—	—	—
Benzene		C	I	P	RR	L	—	—	R	—	R
Benzidine		C	S	—	RR	L	—	—	R	—	R
Benzo(b)fluoranthene			S	—	RR	—	—	—	—	—	R
Benzo(a)pyrene			S	C	RR	—	—	—	—	—	—
Beryllium		PC	S	R	RR	—	—	—	—	—	—
Beryllium compounds		PC	S	—	RR	—	—	—	—	—	—
Bis(2-chloroethyl)ether (BCEE)			L	—	RR	L	—	—	—	—	—
Bis(chloromethyl)ether (BCME)		C	S	—	RR	L	—	—	R	—	—
Cadmium		PC	S	C	RR	—	R[a]	R	—	—	—
Cadmium compounds			I	—	RR	L	—	R	—	—	—
Carbon tetrachloride			S	C	RR	L	—	—	—	—	—
Chlordane	S		L	—	RR	L	—	R	—	—	—
Chlorobenzilate	S		L	—	—	L	—	R	—	—	—
Chloroform (a trihalomethane, THM)	S		S	C	RR	L	R	—	—	—	—
Chloromethyl ether		C	S	—	—	—	—	—	R	—	—

Substance	1	2	3	4	5	6	7	8	9	10	11
Chromium compounds (hexavalent)	—	—	—	—	—	L	RR	—	S	C	—
Coal tar and soot	—	—	—	—	—	—	—	—	S	C	—
Coke oven emissions (polycyclic organic matter; POM)	—	—	R	—	—	—	—	C	N	—	—
Creosote	—	—	—	—	—	—	—	—	N	—	—
Cyclamates	—	—	—	—	—	—	—	—	N	—	—
D&C Blue No. 6	—	R	—	—	—	—	—	—	N	—	—
D&C Red No. 10	—	R	—	—	—	—	—	—	N	—	—
D&C Red No. 11	—	R	—	—	—	—	—	—	N	—	—
D&C Red No. 12	—	R	—	—	—	—	—	—	N	—	—
D&C Red No. 13	—	R	—	—	—	—	—	—	N	—	—
D&C Yellow No. 1	—	R	—	—	—	—	—	—	N	—	—
D&C Yellow No. 9	—	R	—	—	—	—	—	—	N	—	—
D&C Yellow No. 10	—	R	—	—	—	—	—	—	N	—	—
DDT (dichlorodiphenyltrichloroethane)	—	—	—	R	—	L	RR	—	L	—	—
Dibenz(a,h)anthracene	—	—	—	R^a	—	—	RR	—	S	—	—
1,2-dibromo-3-chloropropane	—	—	R	—	—	L	RR	C	S	—	S
1,2-dibromoethane	—	R	—	—	—	L	RR	—	N	—	—
3,3'-dichlorobenzidine	—	—	—	R	—	L	RR	—	S	—	S
1,2-dichloroethane	—	R	—	—	—	L	RR	C	S	—	I
Dieldrin	—	—	—	—	—	—	—	—	L	—	—
Diethylpyrocarbonate	—	—	—	—	—	—	—	—	N	—	—
Diethylstilbestrol (DES)	—	—	R	—	—	L	RR	—	S	C	—
4-dimethylaminoazobenzene	—	—	—	R	—	—	RR	—	S	—	—
2,4-dinitrotoluene	—	—	—	—	—	—	—	—	N	—	—
1,4-dioxane	—	R	—	—	—	L	RR	C	S	—	L
1,2-diphenylhydrazine	—	—	—	R	—	L	RR	—	N	—	—
Dulcin	—	—	—	—	—	—	—	—	L	I	—
Epichlorohydrin	—	—	R	—	—	L	RR	C	L	C	—
Ethylene bis dithiocarbamate	—	—	—	—	—	—	—	—	N	C	—
Ethylene oxide	—	R	—	—	—	—	—	C	I	PC	—
FD&C Red No. 2	—	R	—	—	—	—	—	—	N	—	—
FD&C Violet No. 1	—	R	—	—	—	—	—	—	N	—	—
Formaldehyde	—	—	—	—	—	L	—	C	N	—	—
Graphite	—	R	—	—	—	L	—	—	N	—	—
Heptachlor	—	—	—	R	—	L	RR	—	L	—	(S)

Table 8-4 continued

Chemical	Evaluation by — NCI — A	IARC H	IARC A	Statutes — CAA	CWA §307	CWA §311	SDWA	FIFRA	OSHA	FDCA	CPSA
Hexachlorobenzene	—	—	S	—	RR	L	—	—	—	—	—
Hexachlorobutadiene	—	—	N	—	RR	L	—	—	—	—	—
Hexachlorocyclohexane	—	I	L	—	—	L	—	—	—	—	—
α-hexachlorocyclohexane	—	—	N	—	RR	L	—	—	—	—	—
β-hexachlorocyclohexane	—	—	N	—	RR	L	—	—	—	—	—
Hexachloroethane	(S)	—	N	—	RR	L	—	—	—	—	—
Ideno(1,2,3-cd)pyrene	—	—	S	—	RR	—	—	—	—	—	—
Kepone (chlordecone)	S	—	S	—	—	L	—	V	—	—	—
Lindane	—	—	L	—	RR	L	R[a]	R	—	—	—
Mercaptoimidazoline	—	—	N	—	—	—	—	—	—	R	—
4,4' methylene bis (2-chloroaniline)	—	—	S	—	—	—	—	—	—	R	—
α-naphtylamine	—	—	L	—	—	—	—	—	R	—	—
2-naphthylamine	—	C	S	—	—	—	—	—	R	—	—
Nickel	—	PC	S	C	RR	—	—	—	—	—	—
Nickel compounds	—	PC	S	—	RR	L	—	—	—	—	—
Nitrosamines	—	—	N	—	—	—	—	—	—	—	—
4-nitrobiphenyl	—	—	L	—	—	—	—	—	R	R	—
N-nitrosodi-n-butylamine	—	—	S	—	—	L	—	—	—	—	—
N-nitrosodiethylamine (DENA)	—	—	S	—	—	L	—	—	—	—	—
N-nitrosodimethylamine (DMNA)	—	—	S	C	RR	L	—	—	R	—	—
N-nitrosodi-n-propylamine	—	—	S	C	RR	—	—	—	—	—	—
N-nitroso-N-ethylurea (NEU)	—	—	S	C	—	—	—	—	—	—	—
N-nitroso-N-methylurea (NMU)	—	—	S	C	—	—	—	—	—	—	—
Oil of calamus	—	—	N	—	—	—	—	—	—	R	—
P-4000	—	—	N	—	—	—	—	—	—	R	—
Pentachloronitrobenzene (PCNB)	I	—	N	—	—	—	—	R	—	—	—
Polychlorinated byphenyls (PCBs; Toxic Substances Control Act-RR)	—	I	S	C	RR	L	—	—	R	R	—
β-propiolactone	—	—	S	—	—	—	—	—	—	—	—

Substance	NCI	IARC	CAA	CWA	SDWA	FIFRA	OSHA	FDCA	CPSA
Safrole	S	—	—	—	—	—	—	—	—
2,3,7,8-tetrachlorodibenzo-p-dioxin (TCDD, dioxin)	N	—	C	—	RR	—	—	—	—
1,1,2,2-tetrachloroethane	N	(S)	C	L	RR	—	—	—	—
Tetrachloroethylene (perchloroethylene)	N	(S)	C	L	RR	—	—	—	—
Thiourea	S	S	—	—	—	R^a	—	—	—
Toxaphene	S	S	—	L	RR	R	—	—	—
1,1,2-trichloroethane	N	(S)	C	L	RR	—	—	—	—
Trichloroethylene	L	(S)	C	L	RR	—	—	—	—
2,4,6-trichlorophenol	S	S	—	L	RR	R	—	—	—
Trihalomethanes (THM)	N	—	—	—	—	—	—	—	—
TRIS (flame retardant)	N	—	—	—	—	—	—	—	R
Vinyl chloride	S	C	R	L	RR	—	R	V	R
Vinylidene chloride	N	—	C	L	RR	—	—	—	—
Radionuclides	—	—	R	—	—	—	—	—	—

Source: Office of Technology Assessment.

Abbreviations:

NCI National Cancer Institute data
IARC International Agency for Research on Cancer evaluation
A = Animal evidence
S = Sufficient evidence for carcinogenicity (for more description see chapter 4, appendix A)
(S) = Class 3 of NCI; very strong evidence in 1 species; no evidence in 2nd species
L = Limited evidence for carcinogenicity
I = Inadequate evidence for carcinogenicity
H = Human evidence
C = Identified as a carcinogen from human studies
PC = Identified as a probable carcinogen from human studies
I = Inadequate evidence to reach a conclusion about carcinogenicity from human studies
N = Not evaluated

CAA Clean Air Act
CWA Clean Water Act
SDWA Safe Drinking Water Act
FIFRA Federal Insecticide, Fungicide, and Rodenticide Act
OSHA Occupational Safety and Health Act
FDCA Food, Drug, and Cosmetic Act
CPSA Consumer Product Safety Act
C = Being considered for regulation.
P = Regulation proposed
R = Regulated
RR = Regulation required by act
L = Discharge levels restricted
V = Voluntarily withdrawn from market

^{a}Regulation based on noncarcinogenic toxicity (in addition to those indicated, many other listed substances encountered in the work place are regulated because of toxicities other than carcinogenicity).

The absence of an entry under NCI or IARC does not necessarily mean that the evidence about carcinogenicity is poor or limited. For instance, although the first substance, 2-acetylaminofluorene, does not occur on either the NCI or the IARC list, it is an accepted animal carcinogen. Furthermore, other chemicals have been reviewed since the IACR publication, but the results of the reviews are not yet available.[27]

Some complexities of regulating carcinogens are demonstrated by the table. Some substances present a risk in locations covered by different laws, and separate regulations are necessary for each exposure. Under the Clean Air Act, the EPA has proposed regulation for, or regulated, six carcinogens and is considering an additional twenty-four. Section 311 of the Clean Water Act deals with oil and hazardous spills and is not focused on regulating carcinogens, but hazardous-discharge reporting levels have been promulgated for the listed chemicals, and carcinogenicity was considered in setting those levels. The forty-nine substances for which regulation is required under Section 307 of the act were included in the law in 1977. Standards have been set for trihalomethanes, including chloroform, under the Safe Drinking Water Act. A few metals and pesticides that are identified as carcinogens are regulated under the same act but because of other toxic properties. Implementation of the FIFRA has resulted in voluntary withdrawal of pesticides before regulations were promulgated as well as regulations restricting or forbidding use.

The OSHA has regulated the substances shown because of carcinogenicity, and many other substances on the list are regulated in the work place because of other toxicities. FDA regulation of carcinogenic food additives and colors has eliminated most of the listed colors and sweeteners from the food supply. The Consumer Product Safety Commission has regulated five chemicals and benzidine-containing dyes.

The table does not discriminate between regulations that set a permissible limit, such as the OSHA standards, and those that ban a substance, such as FDA regulations of food colors. The entry R indicates only that some regulation is in effect.

The laws are designed to reduce exposure to carcinogens. They may regulate too many or too few chemicals, but chemicals are being regulated. Furthermore, apparently, few nonrisky chemicals have been regulated under the current system.

Notes

1. R. Rettig, *Cancer Crusade: The Story of the National Cancer Act of 1971* (Princeton, N.J.: Princeton University Press, 1977).

2. Union Carbide, *The Vital Consensus: American Attitudes on Economic Growth* (New York: Corporate Communications Department, 1980).

3. Regulatory Council, *Regulation of Chemical Carcinogens* (Washington, D.C., 1979).

4. U.S. Congress, Office of Technology Assessment, *Cancer Testing Technology and Saccharin* (Washington, D.C. Government Printing Office, 1977).

5. Ibid.; and National Research Council/Institute of Medicine, *Food Safety Policy: Scientific and Societal Considerations* (Washington, D.C.: National Academy of Sciences, 1979).

6. U.S. Congress, House, Committee on Interstate and Foreign Commerce, *Report on the Toxic Substances Control Act,* Rep. No. 94–1341, 94th Cong., 2d sess., 1976.

7. H.P. Green, "The Role of Law in Determining Acceptability of Risk," in *Management of Assessed Risk for Carcinogens,* W.J. Nicholson, ed. (New York: New York Academy of Sciences, 1981).

8. A.M. Freeman, III., "Technology-Based Effluent Standards: The U.S. Case," *Water Resources Research* 16 (1980):21–27 (also available as RFF Reprint 178 from Resources for the Future, Washington, D.C.).

9. D. Doniger, Natural Resources Defense Council, personal communication, March 1981.

10. Food and Drug Administration, "Polychlorinated Biphenyls (PCB's); Reduction of Tolerances, Final Rule," *Federal Register* 44 (1979): 37336–37403.

11. Lester Lave, *The Strategy of Social Regulation* (Washington, D.C.: The Brookings Institution, 1981).

12. A.M. Freeman, III, Bowdoin College, personal communication, March 1981.

13. J.C. Davies, S. Gusman, and F. Irvin, *Determining Unreasonable Risk under the Toxic Substances Control Act* (Washington, D.C.: Conservation Foundation, 1979).

14. National Research Council, *Principles for Evaluating Chemicals in the Environment* (Washington, D.C.: National Academy of Sciences, 1975).

15. Ibid.

16. Ibid.; and Davies, Gusman, and Irvin, *Determining Unreasonable Risk.*

17. H.T. Markey, "Statement" U.S. Congress, House, Comm. *Risk/Benefit Analysis in the Legislative Process,* Subcommittee on Science, Research, and Technology, U.S. House of Representatives, 96th Cong., 1st sess., 1979.

18. R. Field, "Statutory Language and Risk Management" (Report

prepared for the Committee on Risk and Decision Making, National Research Council, Washington, D.C., 1980).

19. M.S. Baram, "Regulation of Health, Safety, and Environmental Quality and the Use of Cost-Benefit Analysis" (Final report to the Administrative Conference of the United States, Washington, D.C., 1979).

20. Markey, *Risk-Benefit Analysis.*

21. Field, "Statutory Language."

22. Ibid.

23. Ibid.

24. R.A. Griesemer, and Cueto, C., Jr., "Toward a Classification Scheme for Degrees of Experimental Evidence for the Carcinogenicity of Chemicals for Animals," in *Molecular and Cellular Aspects of Carcinogen Screening Tests,* R. Montesano, H. Bartsch, and L. Tomatis, eds. (Lyons, France: International Agency for Research on Cancer, 1980).

25. International Agency for Research on Cancer, *Chemicals and Industrial Processes Associated with Cancer in Humans,* IARC Monographs, Supp. 1 (Lyons, France, 1979); and IARC, "An Evaluation of Chemicals and Industrial Processes Associated with Cancer in Humans Based on Human and Animal Data," IARC Monographs, vol. 1 to 20, *Cancer Research* 40 (1980):1-20.

26. See, for example, International Agency for Research on Cancer, ibid.

27. Ibid.

To Breathe Clean Air

National Commission on Air Quality

Standards to Protect the Public Health

The effects of technological advances on human health are only beginning to be understood. Humans are exposed to potentially harmful substances in a variety of ways such as through air and water and food, drugs, and other products they use. Adverse health effects range from temporary ones such as eye, throat, and chest irritations and coughs to chronic effects such as lung diseases, cancers, and other problems resulting from prolonged exposure to low levels of a pollutant. Evidence relating a particular pollutant to specific effects often is limited and likely to be suggestive rather than conclusive. However, when the health of potentially large numbers of people could be affected or when the health of those exposed to a pollutant may be severely impaired, the prudent public policy is to act early to ensure that the public's exposure to the suspected hazard is limited. For pollutants suspected of causing cancer and other chronic disorders, the time between exposure and discernible effects may be decades. In those cases, regulatory decisions, particularly for new toxic substances, almost certainly will have to be made in the absence of definitive evidence.

Congressional Action

Congress based the Clean Air Act on the principle that to protect public health, government must act to control potentially harmful pollutants despite scientific uncertainty about the precise harm they cause and the levels of exposure that cause that harm. The act required the EPA to set national primary air-quality standards to protect against pollution levels that "endanger public health, allowing an adequate margin of safety."

Section 109 of the act required the EPA to set national air-quality standards for air pollutants that are widely dispersed and emitted by large numbers or a wide variety of sources. The act does not attempt to specify what constitutes the adverse health effects against which the standards must

This chapter is extracted from the report of the National Commission on Air Quality, *To Breathe Clean Air* (Washington, D.C.: U.S. Government Printing Office, 1981).

protect. Some commentators have suggested that greater specificity is needed.[1] However, any statutory attempt to identify specific types of health effects could inhibit identification of unanticipated effects that should be considered. In addition, most air pollutants affect public health in more than one way. To carry out its responsibilities under the act, the EPA, before making a final decision to set or revise an air-quality standard, must have all relevant and reliable scientific information on the full range of possible health effects of a pollutant.

The statutory basis for setting national primary air-quality standards does not take economic factors into account. In the act, the Congress recognized that, while the levels of air pollution at which the public health is affected usually do not vary among different locations, the costs of meeting a specific standard can vary substantially from area to area, depending upon the severity of the pollution. Thus, if a national air-quality standard were based in part on the costs of complying with it, the high costs of meeting the standard in a few heavily polluted areas could result in the standard's being set at a less-protective level than is achievable in a reasonable, economic fashion in other areas. The health benefits of good air quality and the economic, social, energy, and other costs of meeting health-based standards can be advanced more effectively and appropriately when control programs are established for particular areas than when national primary standards are set.

The act provided avenues of regulation other than Section 109 for pollutants more localized in origin than those requiring national ambient standards. Section 111 authorizes the EPA to establish new-source performance standards and to require that states establish performance standards for existing sources of pollutants that "may reasonably be anticipated to endanger public health or welfare." Under Section 112, the EPA must establish national emission standards that provide an "ample margin of safety" for especially hazardous air pollutants that may "result in an increase in mortality or an increase in serious irreversible, or incapacitating reversible, illness."

Like the language of Section 109, portions of Section 112 have raised questions of interpretation. The requirement of allowing an ample margin of safety in setting standards presumes a threshold concentration below which there is no adverse health effect. For pollutants that cause or are suspected of causing cancer, it is currently assumed that any level of exposure may be harmful.

In 1977, the Congress recognized the need to improve the EPA's ability to sustain its decisions on specific standards against legal challenge.[2] The 1977 amendments to the act specified that decisions regarding identification of pollutants and levels for standards were to be made "in the judgment of the administrator." With this language, Congress made explicit what was contemplated in the 1970 act—that the EPA is delegated substantial authority to weigh risks and make decisions on questions on the frontiers of scientific knowledge. Recognizing the need to improve and refine the scientific bases for national air-quality standards, Congress in 1977 also required the administrator to review criteria and standards at five-year intervals.

While delegating substantial authority to the EPA to act despite scientific uncertainty, Congress also acknowledged criticisms of the soundness of the EPA's research. Criticism of the scientific bases used by the EPA in setting standards under Section 109 led Congress in 1977 to establish a seven-member, independent scientific committee, known as the Clean Air Scientific Advisory Committee (CASAC) Every five years, CASAC is to review criteria published under Section 108 and standards promulgated under Section 109 and to recommend any revisions it finds appropriate. The 1977 amendments also required the EPA to publish, with any proposed standard under Sections 109, 111, or 112, the CASAC's comments on the standards and the reason for any differences between the comments and the EPA's basis for the proposed standard.

In addition, in the Environmental Research, Development and Demonstration Authorization Act of 1978, Congress gave statutory recognition to the Science Advisory Board (SAB), which the EPA had created in 1974, and required the agency to make available to the board, for discretionary scientific and technical review any proposed criteria document, standard, limitation or regulation under the Clean Air Act.

EPA Activities

Since 1970, the EPA has issued national ambient air-quality standards under Section 109 for seven pollutants: sulfur dioxide, carbon monoxide, total suspended particulates, photochemical oxidants, hydrocarbons, nitrogen dioxide, and lead. The first six were required by the act; lead was regulated as the result of a lawsuit brought by the Natural Resources Defense Council in 1975.

The process for establishing national ambient air-quality standards is lengthy and complex (see figure 9–1). As part of the process, the law requires the EPA to prepare a criteria document that provides the scientific basis for the standard. Under current procedures, an office under the assistant administrator for R&D prepared a draft document for review by the CASAC and release to the public. While the criteria document is being developed, a division under the assistant administrator for air, noise, and radiation (air-program office) prepares a staff paper, which is a summary of the evidence and conclusions in the criteria document and of their significance for the regulatory decision. Recent staff papers have been released

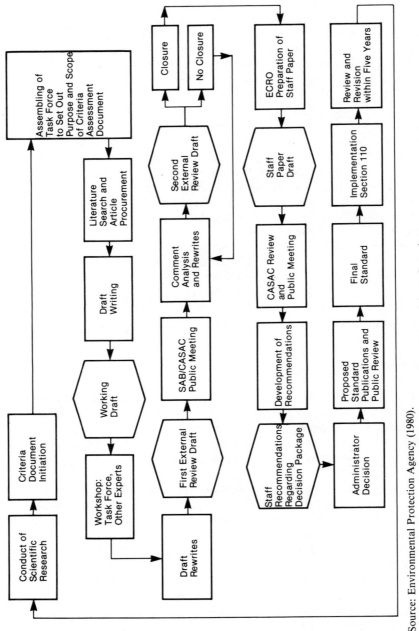

Source: Environmental Protection Agency (1980).

Note: SAB/CASAC = Science Advisory Board/Clean Air Scientific Advisory Committee

CASAC = Clean Air Scientific Advisory Committee

ECRO = Environmental Criteria and Assessment Office

Figure 9–1. Standard-Setting Process at the EPA

to the CASAC for review along with the draft criteria document. A recommended standard is prepared under the authority of the assistant administator for air, noise, and radiation, sent to the administrator for approval, and published as a proposed standard. After a public-comment period, the proposed standard is reviewed and modified, as appropriate, and a final standard is published.

Since 1978, responsibility for overseeing preparation of the criteria document has been assigned to the Environmental Criteria and Assessment Office in the Office of Research and Development. The document, which is intended to focus on objective scientific and engineering data, describes and evaluates these data in terms of quality, completeness, and pertinence to health and welfare. The following subjects are covered:

> Chemical and physical properties of the substance, methods of measurement, pattern of global cycling, geographic and temporal distribution of sources, levels of emissions, and ambient concentrations;

> Health effects, based on evidence from in vitro, animal, clinical, and epidemiological studies; factors influencing toxicity; and populations-at-risk;

> Welfare effects, including effects on plants, animals, ecological systems, and materials.

A task force made up of representatives from regulatory and research offices in the EPA selects authors and reviewers for the document. Primarily scientists and engineers with knowledge of the subjects to be covered, they may come from research offices in the EPA or from outside the agency.

The Environmental Criteria and Assessment Office is relatively new, and its assignment is complex. A major concern expressed by the CASAC and other reviewers is that the criteria documents have become encyclopedias of insufficiently analyzed material, much of which is of poor quality.[3] Despite the statutory directive in Section 108 that requires the documents to include "the latest scientific knowledge useful in indicating the kind and extent of all identifiable effects on public health or welfare," the Environmental Criteria and Assessment Office in the past has included virtually all available data on the subject, some of which has been of questionable value to the regulatory decision. Because the office can influence the emphasis given to specific contents of the document, its occasional insufficient response to certain critical commentary from both the public and its own consultant-reviewers and EPA staff's participation in the authorship of the document have been cited as factors appearing to compromise the objectivity of the document.[4]

Another issue affecting criteria-document development is the extent to which the process must accommodate public participation. The Federal Advisory Committee Act requires that meetings between federal-agency personnel and two or more advisors be announced at least fifteen days in advance and be open to the public. In 1979, the EPA settled a court suit with the American Iron and Steel Institute by acknowledging that workshops convened to allow authors of a criteria document to review and evaluate one another's work were subject to the Federal Advisory Committee Act and must be open to the public. Representatives of the business community have been the primary public participants in the few workshops held since the settlement. The sessions have been recorded, and transcriptions of the proceedings are available to the public.

Opinion is divided as to whether a public forum is conducive to full exploration of issues and to critical review by authors of one another's preliminary work prior to the first external draft review. Confidentiality prior to publication of a first draft is a standard scientific practice that encourages the proposal of new hypotheses and peer review of the accuracy, significance, and implications of data and hypotheses. Opening this preliminary stage of the process to the public has hampered the scientific procedure and could impair the EPA's ability to enlist independent scientists and engineers to develop criteria documents.[5] Some observers point out, however, that the presence of outside parties can encourage careful, precise, and objective authorship. Whether authors and agency staff should meet in closed session during some stage of document development is an unresolved question.

The staff paper prepared by the Office of Air, Noise, and Radiation is intended to translate scientific information in the criteria document into a form that will be useful in determining the level at which to set a standard.

Staff papers on carbon monoxide and nitrogen oxides have been reviewed by the CASAC at its request. Committee members said the papers provided a means of underscoring the important scientific evidence in the criteria documents. Members of the CASAC perceive staff-paper reviews to be a first step in the committee's broader mandate under the act "to advise the administrator of any adverse public health, welfare, social, economic or energy effects which may result from . . . national ambient air quality standards." Opinion is divided within the EPA, among CASAC members, and among outside interests about the desirability of having the CASAC advise the EPA on policy as well as scientific issues.[6]

Another issue concerning the CASAC's role in the criteria-document-review process is the size of its membership. To fulfill their statutory mandate, CASAC members must be competent in a variety of biomedical, geophysical, chemical, mathematical, and engineering disciplines. The CASAC is limited by statute to seven members, including at least one member of the

NAS, one physician, and one person representing state air-pollution-control agencies. Consultants, although used routinely, cannot vote. Other committees of the SAB, which perform functions similar to those of the CASAC, are not statutorily limited in size. A commission study has suggested that limiting CASAC membership to seven may reduce its ability to perform its functions.[7]

Members of the SAB—and therefore of the CASAC, which has been established as a standing committee of the SAB—are appointed by the EPA administrator on the recommendations of interested groups. Many SAB members are affiliated with universities, and a significant number are involved in NAS activities. As intended by Congress, the CASAC appears to be functioning independently of EPA influences and providing detailed, objective assessments of documents prepared by the EPA. Its independent judgment has been demonstrated in its critical reviews of the current generation of criteria documents for ozone, nitrogen dioxide, carbon dioxide, total suspended particulates, and sulfur dioxide. In addition, the CASAC and SAB have performed their functions on a more-rapid and timely schedule than the NAS, which the EPA also calls upon to assess scientific and related issues.

After CASAC review, the staff paper is used to prepare materials for the EPA administrator's consideration in setting a standard. At this stage, the administrator is called upon to act on questions on the frontiers of scientific knowledge. The factual information available in most cases does not provide a complete basis for determining what standard is necessary to protect public health. Although scientists often can identify biochemical or physiological changes in the human organism as a result of exposure to a pollutant, they may not be able to conclude that the change is, or is not, likely to contribute to impairment of people's health. Setting a standard thus is not a matter of reaching a scientific conclusion but is a policy judgment—based on scientific facts, estimates, and hypotheses drawn from still emerging data—of what regulatory action is needed to "prevent harm before it occurs."[8]

The EPA office responsible for preparing standard-setting materials has a program underway to develop, test, and refine quantitative risk-assessment techniques to assist in estimating the extent of uncertainty in the scientific information. The office attempted this approach during revision of the ozone standard and was criticized by industry and by the SAB for using insufficiently developed techniques.[9] Since that time, the agency has initiated a more-extensive program to improve and test methodologies.

Risk assessment typically includes both qualitative and quantitative stages. The detailed, objective review of data is qualitative. The next stage, quantitative assessment, involves establishing the probabilities of various outcomes based on available scientific data. While quantitative risk assessment is complex, speculative, and subject to a number of uncertainties, its use by the EPA in setting air-quality standards encourages formal and rigorous analysis in a process that is inherently complex and can act to refine research and data collection.

Although the EPA is not to consider economic factors in setting a national primary standard, a regulatory analysis that accompanies the materials upon which the agency bases its decision includes an assessment of the economic impact of the proposed standard. An executive order issued in 1978, and superseded by an order issued on 17 February 1981, requires agencies to analyze and publish the possible economic consequences of proposed regulations that will have an estimated annual effect on the economy of at least $100 million. Economic analyses have been conducted as part of the process of revising the standards for ozone, carbon monoxide, sulfur dioxide, and particulates.

Economic Considerations

Since the inception of governmental efforts to control air pollution, Congress and the public have realized that significant economic consequences would result from these efforts. Reducing air pollution would reduce the costs of health care and damages to soils, water, crops, wildlife, materials, visibility, and valued objects, but it also would require significant expenditures to be made for controlling mobile and stationary sources of pollution. Accordingly, federal laws addressing air-pollution problems have reflected the judgment of Congress that an appropriate balance be made between the need to protect against the effects of air pollution and the costs and other economic effects of the protection.

Congress approached the consideration of economic effects under the Clean Air Act in two ways. For several major elements of the federal air-pollution-control program, such as requiring achievement of ambient air-quality standards by fixed deadlines, Congress itself determined that the economic effects of any necessary control activities are acceptable and, therefore, that no explicit consideration of economics by the EPA in establishing standards is necessary. For other major programs under the act, Congress required explicit consideration of economic effects. For new stationary sources, for example, economic considerations have resulted in the best possible controls' not being required.

While acknowledging that there has been some consideration of economic factors in determining air-pollution-control requirements, many critics of the act, particularly industries subject to control requirements, have contended that the benefits of air-pollution control often do not justify the economic consequences and that costs and benefits have not always been appropriately balanced. Accordingly, proposals have been made for closer examination of the relative costs and benefits of all or parts of the act's programs and for the requirement that specific techniques be used in evaluating costs and benefits.

Studies performed to date regarding costs and benefits of air-pollution-control programs do not provide unequivocal conclusions. However, they do permit certain observations:

Both costs and benefits of the act's programs have been significant.

Quantification of all costs and benefits of any air-pollution-control program is virtually impossible.

Necessary data to make reasonably accurate cost-benefit comparisons are not likely to be available in the foreseeable future because a number of relevant variables cannot be easily measured.

Recognition of national economic problems and of the costs of further pollution controls has focused attention on the potential for incorporating economic incentives and approaches into federal clean-air laws and programs. These include concepts such as the bubble policy, emission banking, emissions fees, and tax incentives.

This section presents the economic effects of air-pollution-control requirements as reflected in a number of major economic indicators. The results of recent studies regarding estimates of the aggregate costs and benefits of air-pollution-control requirements and the limitations inherent in these estimates are discussed. The results of studies undertaken by the commission and others of the economic consequences of the Clean Air Act's programs, including the provisions of the act that currently require consideration of economic factors, are presented. Economic-incentives approaches used in the United States and elsewhere to maximize efficiency in meeting pollution-control objectives also are analyzed.

Costs and Benefits: National Effects

The direct and indirect costs and benefits resulting from the act occur throughout the national economy, individual industries, and private households. Although environmental and other government regulations frequently are cited as major factors affecting the pace and direction of economic growth, effects of those regulations, including costs and benefits resulting specifically from air-pollution control, are extremely difficult to measure accurately.

Although the economic effects of all pollution control—including air-pollution control—on national economic indicators such as the gross national product (GNP), consumer price index (CPI), and labor productivity have not been substantial, they nonetheless represent a significant commitment of resources. Estimates of economic effects are based on data for pollution-abatement investments and expenditures after subtracting control costs that would have been incurred regardless of federal regulations. However, the significance of these data for air-pollution control is limited because they also include estimated effects from water-pollution and solid-waste regulations.

The effect of all pollution control on productivity has been an annual average decline of 0.05 of 1 percent between 1967 and 1969; 0.1 of 1 percent between 1969 and 1973; 0.25 of 1 percent from 1973 to 1975; and 0.08 of 1 percent from 1975 to 1978.[10] This latest statistic indicates that gradually less capital and labor are being invested in pollution control.[11]

Productivity and real GNP have fallen and the CPI has risen as a consequence of pollution control, while unemployment has decreased. Labor productivity between 1970 and 1986 is estimated to be declining at an average yearly rate of 0.1 percent, while unemployment is decreasing at the rate of 0.25 percent per year for the period. Pollution-abatement investments contributed to inflation in two ways: (1) increased per-unit cost of goods and (2) higher per-unit labor costs and, thus, higher prices.[12]

Charting the effect of pollution-control requirements on future national economic growth is difficult because of inadequate or unavailable data. In particular, it is difficult to determine the extent to which facilities will or will not be constructed or curtailed solely because of air-pollution-control requirements. A more-productive and -informative means to assess the economic effects of control is to examine them on a smaller scale, such as by individual industry.

Costs of Air-Pollution Control

Estimates of air-pollution-control costs are best expressed as a range of probable expenditures. Although in the aggregate these expenditures represent large absolute amounts, they do not result in a substantial adverse effect nationwide. However, this does not mean that the effect on particular industries or firms may not be significant.

The total annual costs for air-pollution controls for 1978, including operation and maintenance expenditures, are estimated to have been 16.6 billion. Of this amount, $7.5 billion was for capital expenditures and $9.1 billion was for operational and maintenance expenditures. The total annual costs for 1987 are projected to be $37.4 billion (1978 dollars).[13]

Estimates of the costs of capital expenditures related to air-pollution control for 1978 range from $1.8 billion to $7.5 billion.[14] This range represents what industry spent for only air-pollution control on an annualized basis. Estimates of 1978 capital expenditures that have not been annualized range between \$3.6 billion and \$3.8 billion.[15]

In 1978, all industries spent an average of approximately 2.38 percent of total capital expenditures on air-pollution control. The steel industry incurred the highest expenditure—11.06 percent of all building and plant costs. The lowest was incurred by the railroad industry, which expended 0.19 percent of total plant and equipment costs.[16]

Estimates of nonannualized expenditures for air-pollution control in 1979 range from \$3.9 billion to \$4.4 billion.[17] Total industry expenditures for 1979 for air-pollution control were projected to be 2.3 percent of all plant and equipment expenditures. Again, the steel industry experienced the highest percentage of air-pollution-control costs—11.93 percent.[18]

The size of operations, local conditions, and the nature of production and pollution-control systems make costs vary around the country, by industry, and even within industries. The Bureau of Economic Analysis survey reports that the highest cost (\$2 billion) in 1978 was incurred by the manufacturing sector, which included primary metals, petroleum refining, and all other manufacturing processes. Public utilities followed, with \$1.4 billion.[19]

Capital costs present one of the best available indicators of the costs of clean air, although they are by no means comprehensive. Air-pollution control also imposes indirect costs, but these costs are much more elusive. Indirect costs include the loss of income that is diverted from production and spent on air-pollution control, costs associated with record keeping and regulation requirements, and expenditures for operation and maintenance of control facilities and equipment.

Further difficulties in estimating national or industry costs are caused by the lack of uniformity in methods of determining costs. Although the estimates reported here are comparable, adjustments have been made so that they account for the same economic variables. The variables are the effects of depreciation and amortization, extrapolation of air-pollution-control costs from other control expenditures through similar techniques, separating incremental air-pollution-control costs from normal production costs, separating air-pollution-control costs from pollution costs that would have been incurred without the act, and estimating the usual discrepancy between actual and projected costs.

Benefits of Air-Pollution Control

The economic benefits of pollution control are neither readily defined nor easily quantified. Often, benefits accrue in noneconomic terms and must be converted into monetary values. Because of the uncertainties associated with data and the conversion of effects to monetary values, benefit estimates are best reported as approximations and ranges instead of absolute numbers.

The range of economic benefits for air-pollution control reported in one study is from \$4.6 billion to \$51.2 billion.[20] The reported estimates are based on studies that were adjusted on a consistent basis to make them comparable.[21] This range includes improvement to human health, reduced soiling and cleaning costs for households, and reduced damages to vegetation and crops and to materials. These estimates do not attempt to qualify for aesthetic benefits or savings gained by avoiding pollution damage.

Eighty percent of the benefits in this range are attributable to health improvements—that is, fewer illnesses and premature deaths. The basic assumption for the overall range of benefits reported in Freeman's study is a 20 percent reduction in two criteria pollutants: total suspended particulates and sulfur dioxide. The reductions are estimated by Freeman for 1970 to 1977 for particulates and for 1972 to 1977 for sulfur dioxide. Freeman used the EPA's "National Air Quality and Emissions Trends" reports for estimating the average reduction. Other estimates assume greater reductions for these pollutants.[22]

Health Benefits

Health benefits alone comprise the majority of total benefits estimated from air-pollution control. The range of probable values of health benefits can be seen in the amounts reported in several studies. One study reported annual national benefits to be \$16 billion for a 60 percent reduction in air pollutants in urban areas with a total population of 150 million. For purposes of quantification, an individual life was assigned a value of \$1 million in this study.[23]

A different estimate that considered worker productivity—that is, the effects of reduced absenteeism and illness—produced yearly national-health benefits of \$43 billion.[24] For studies that took into account socioeconomic and life-style factors such as diet and smoking, health benefits varied from \$3 billion to \$40 billion annually for a 20 percent reduction of air-pollution levels between 1970 and 1978.[25] For studies using factors such as medical costs, loss of earnings due to illness, costs of temporary household help required during illness, and other, less-significant costs, the range of annual benefits is between \$3 billion and \$43 billion.

Nonhealth Benefits

Benefits can also be measured as gains from avoiding expenses caused by pollution. Benefits from costs avoided result from reduced pollution affecting vegetation, commercial crops, and materials exposed to the air. To obtain estimates of these avoided costs, the commission undertook a study of nonhealth benefits.

The study found the approximate value of reduced damage to nineteen food and fiber commercial crops from meeting secondary air-quality standards in 1978 to be $1.78 billion (1980 dollars); the total market value of these crops in 1978 was $54 billion (1980 dollars). This estimate represents the effect on these crops of lowering levels of sulfur dioxide and ozone throughout the country. It does not include the value of reduced damage to other crops or to timber and ornamental plants.[26]

For the crops examined in the study, ozone causes 98 percent of the reported damage. Crops usually suffer slightly more from short-term ozone exposure than from multiday exposure. The states hardest hit by pollution damage to crops are Ohio, Indiana, Illinois, and Wisconsin. This region reports approximately 41 percent ($731 million) of all crop damage. The Pacific region (Washington, Oregon, California, Alaska, and Hawaii) is the next most affected, with $384 million in pollution damages, or about 22 percent of all crop damage. Crops in the mountain region (Montana, Idaho, Nevada, Wyoming, Utah, Colorado, Arizona, and New Mexico) are least affected by pollution damage.[27]

The commission study found the value of damage avoided by meeting secondary air-quality standards for the selected metals, fabrics, building materials, rubber, and plastics to be $3.95 billion in 1978. The effects of reducing three pollutants—total suspended particulates, ozone, and sulfur dioxide—form the basis of this estimate. Reductions of sulfur dioxide and ozone, whose corrosive effects on materials cannot be separated, create the largest savings of costs avoided: $3.824 billion, or almost 97 percent. Of the building materials selected, total suspended particulates damage only concrete; reducing this pollutant produces a $113 million benefit.[28]

Avoided damages to individual materials range from $333 per short ton of tin to $4.10 per short ton of aluminum. Nickel shows pollution-damage costs avoided of $304 per short ton, while the damages to plastics and copper are estimated to be only $15 and $18.90 per short ton respectively. For rubber products, pollution particularly affects the value of medical and house products, reducing their service life by about 50 percent.[29]

Fabrics and clothes are most adversely affected by total suspended particulates. The benefits to individuals and households from reducing this pollutant are estimated to be savings of $535 million. The largest savings benefit, $130 million, is in the mid-Atlantic region (New York, Pennsylvania, and New Jersey), followed by the eastern part of the north-central region (Wisconsin, Michigan, Illinois, Indiana, and Ohio), with $118 million.[30]

In areas with high levels of motor-vehicle emissions, property values tend to be lower and wages higher to compensate for the polluted surroundings. Consequently, reducing emissions results in economic benefits. Reductions in nitrogen oxides and photochemical oxidants have been estimated to create property-value improvements of between $1.5 billion and $5 billion nationally and estimated wage benefits of $5 billion.[31] (The same study estimated that when the benefits owing to reduced damage to materials, health, and vegetation from a 35 percent to 80 percent decrease in mobile-source emissions are added to the benefits calculated by the property values or wage-rate differentials, a range of values from $2.5 billion to $10 billion is derived.)

Although good visibility is valued for aesthetic reasons, its value also can be estimated in terms of individuals' willingness to pay for cleaner air and improved visibility. According to national estimates derived from a survey in the western United States, the general population is willing to pay $0.5 billion for improved visibility.[32] A study in the Four Corners region (Arizona, New Mexico, Colorado, and Utah) that asked individuals what they would pay to improve visibility and to live in an area with naturally clean air produced estimates of $15.5 million and $24.6 million, respectively, for the value of the two benefits.[33] Residents in the Los Angeles area reported a willingness to pay $29 per month per household for a 30 percent improvement in air quality.[34]

Limitations of Cost-Benefit Estimates and Analyses

The preceding discussion highlighted wide variations among estimates of costs and benefits reported in numerous studies. Any estimate of expenditures and benefits related to requirements of the act depends upon numerous subjective variables used in its calculation. Every assumption and body of data contain limitations and qualifications that make the development of unchallengeable estimates virtually impossible.

The same data, in fact, can produce contrary results. Estimates that use different assumptions for pollution-reduction levels, current prices, and the monetary value of human health will arrive at different conclusions. For instance, a recent study of air-quality costs and benefits analyzed data used in earlier studies. However, by making different assumptions, the later study estimated benefits as only 20–25 percent of those reported in the earlier study.[35]

A report on methods for estimating costs prepared for the commission noted a number of limitations with these estimates, including the unavailability of complete data bases and the inexact nature of future projections.[36] In addition, any attempt to isolate air-pollution control costs from other costs associated with environmental controls requires that judgments be made about the amount each contributes to the reported total. For example, overhead and operation and maintenance costs must be distributed among air-pollution-control, water-pollution-control, and solid-waste-disposal requirements. Also, different accounting considerations among firms with differing assumptions regarding interest rates, methods of depreciation, and definitions of fiscal year lead to dissimilar reporting of cost estimates.

An additional cause of variation in cost estimates is that some firms may be reluctant to disclose fully cost information for reasons of confidentiality or competition.[37] Projections of future costs also may be inaccurate because advances in pollution-control equipment or production processes are made that were not anticipated at the time the projection was made or because expected advances do not occur. Finally, the sources that gather cost data may define industrial groupings differently, making useful comparisons difficult.

The many limitations associated with the methodology of estimating benefits were analyzed by a commission panel. These limitations include calculating health benefits, which entails applying and analyzing data on ambient pollution levels; correlating human health and air pollutants; and calculating the costs of morbidity and mortality. In addition, health statistics for the general population are incomplete and in some cases unavailable. Placing a monetary value on these statistics generates further problems.[38]

Projecting anticipated benefits to crops and materials requires predicting production levels, prices, and interest rates. Each of these predictions contains considerable uncertainty. Another category of benefits, the aesthetic value of clean air, contains even greater unknowns and degrees of subjectivity. Although clean air is recognized as important, its monetary value eludes systematic, objective quantification. Surveys and questionnaires elicit different responses and result in different estimates. Consequently, the benefits of the act requirements appear to be considerable, yet impossible to measure exactly.

Because of the limitations of the data and the numerous assumptions used in these studies to calculate the costs and benefits of air quality, the commission did not make direct comparisons between the estimated costs and benefits. Both are clearly substantial, but no sound basis exists for stating that one is greater than the other.

One of the most frequently discussed proposals for consideration of economic factors under the act is cost-benefit analysis. This analytical technique provides a means for government regulations and other actions. Recognizing that government discussions often require a reconciliation of competing interests, cost-benefit analysis seeks to quantify and compare costs and benefits of the proposed actions to maximize the efficiency in monetary terms.

The earlier discussion of benefit estimations describes some of the difficulties faced in accurately quantifying both costs and benefits. In addition to the difficulties of quantification, other more-fundamental issues arise if cost-benefit analysis is proposed to be used in evaluating the goals of the act or in establishing health-based standards.

A major limitation is that cost-benefit analysis only addresses economic efficiency. Although it provides the best approach for comparing costs and benefits in monetary terms, it normally does not allow adequate consideration of benefits such as life and health, which are not susceptible to valuation in those terms. If a value is determined by one's willingness to pay, each individual's preference is weighted by the person's relative wealth—that is, by the number of dollars he or she is able to put on that preference. Finally, reliance on the present market to measure the value of longer-term benefits (for example, illness avoided) discounts the value of these benefits in the future.

Finally, use of cost-benefit analysis in establishing national-health-based standards could result in standards being established at levels that could be justified for areas of the country with the severest pollution problems where the cost of controls would be greatest. For example, Los Angeles may not be able to meet the current ozone air-quality standard without imposing measures that would cause severe economic disruption. If such costs are included in establishing a national standard, the cost portion of the cost-benefit analysis would be increased and could lead to the establishment of a national standard that would be acceptable in less severely affected areas. These areas, therefore, could have substantially greater levels of air pollution than would have been allowed if a regional cost-benefit analysis had been performed.

Thus, while cost-benefit analysis may be useful in evaluating certain decisions, its utility is severely limited as a means of establishing a national-health-based air-quality standard.

Consideration of Economics in the Clean Air Act

When Congress passed the Clean Air Act, it recognized that there would be substantial economic consequences of some requirements. Congress established two approaches to economic considerations. It implicitly made a determination that the economic effects of achieving the goals of certain programs were acceptable, and it provided for economic consequences to be evaluated in carrying out other programs under the act.

For several major programs, Congress itself determined that the economic effects of any necessary actions to meet the goals of those programs were acceptable. These programs include the establishment of national ambient air-quality standards and achievement of these standards by established deadlines, the establishment of national emission standards for hazardous air pollutants, and certain aspects of the federal motor-vehicle-control program. The programs for which Congress explicitly ordered the EPA to consider economic effects include the establishment of standards of performance for new or modified sources, development of control requirements in state implementation plans for existing sources in areas that will attain ambient air-quality standards, determination of best-available-control technology for new sources locating in areas achieving the ambient standards, and the remaining sections of the federal motor-vehicle-control program.

Although the 1977 amendments continued the previous requirements for establishing national ambient air-quality standards without consideration of the costs of achieving them by fixed deadlines, Congress took several actions that nonetheless resulted in the economic consequences of these requirements' being taken into account. First, because it was apparent in 1977 that a number of areas could not attain the standards through the adoption of reasonable control measures, and because Section 110 of the 1970 act required that no new or modified sources could be constructed in areas not achieving the standards, Congress added provisions extending the deadline for attainment of standards to 1982 and, in certain instances (for ozone and carbon monoxide), to 1987. Congress believed that these deadlines would provide sufficient time for most areas to achieve the standards through use of reasonable control measures. To facilitate construction of new or modified facilities in these areas before attainment of the standards, Congress added provisions to ensure that facilities could continue to locate in areas not attaining standards.

One of the conditions of new-source construction or modification in areas not meeting standards is that sources apply control technology providing the lowest achievable emission rate. This condition requires use of the most stringent control system available. Once a determination of an achievable emissions rate is made for a particular facility, the costs of meeting the emission requirements are not to be considered; however, commission studies indicate that in many cases the determination of the lowest achievable emission rate is equivalent to new-source performance standards, which take costs into consideration.[39]

In order to permit areas to accommodate economic growth while moving toward attainment of air-quality standards, Congress provided that new sources could locate and that existing sources could expand if the additional emissions were offset by reductions from existing sources. Most new or expanded sources have been able to obtain offsets, although in a small percentage of cases, offsets may not be readily available to sources not already located in an area not achieving ambient standards.

The act's requirement for areas not attaining standards to adopt reasonably available control measures can be viewed as a means of ensuring that no measures that will severely affect an area's economy will be imposed. Although the act states that the controls are to be reasonably available, certain industries or sources within an industry may nonetheless experience substantial economic effects from their imposition.

Since enactment of the 1977 amendments, it has become clear that some areas will not be able to achieve standards by the required dates even with deadline extensions and lowest achievable emission-rate controls, use of offsets, and application of reasonably available control measures. The 1982 deadlines for total suspended-particulate matter will not be achieved in portions of twenty-seven areas, and the 1987 deadline for carbon monoxide and ozone probably will not be achieved in thirty-two counties. A commission study of the New York/metropolitan-Connecticut area projects that after imposition of all reasonable available control measures, an additional 25 percent reduction in emission (which represents the amount of remaining emissions from all vehicle traffic) would not be sufficient to demonstrate attainment of the ozone standard by the 1987 deadline.

The establishment of emission standards for hazardous air pollutants is another provision in which Congress determined that the protection of public health was sufficiently important to warrant the costs of control. The EPA's concern that such requirements for hazardous substances could lead to a zero-emission standard with potentially serious economic effects has been one of the factors leading to the development of the EPA's airborne-carcinogen policy. (No emission standards imposed thus far have required zero emissions.) This policy attempts to consider risks and, at certain points in the process, explicitly considers economic factors.

The emission-reduction requirements for gasoline automobiles were established by Congress in the 1970 act. These required a 90 percent reduction in hydrocarbon and carbon-monoxide emissions by the 1975 model year and a 90 percent reduction in nitrogen-oxides emission by the 1976 model year. The EPA was not allowed to alter the final standard levels; it was authorized to extend the deadline for achievement of these standards for one year. However, in 1974, and again in 1976 and 1977, the automobile manufacturers presented information to Congress regarding the technological, fuel-economy, and economic consequences of meeting the 90 percent requirement by the prescribed deadlines. In 1974 Congress extended the deadline for achievement of the 1975-model-year standard to model-year 1978. In 1977, Congress extended the deadline to model-year 1981 and increased the standard for nitrogen oxides for 0.4 gram per mile to 1 gram per mile.

Other provisions of the act, unlike those discussed, require consideration of economic factors. Section 111 of the act, which requires the establishment of standards of performance for new or modified sources, directs the EPA to consider costs and energy effects as it adopts the standards. The EPA routinely performs an economic analysis of standards under consideration as a part of the process leading to the establishment of final standards. This analysis includes consideration of, among other things, the capital and annualized operating costs of required control technology, the resulting effect on product price, incremental capital expenditure, and rate of return on investment.

The determination of best-available-control technology under the prevention-of-significant-deterioration program for new or modified sources requires the control agency to consider cost, economic, and energy effects in establishing the emission limitation for a source [§169(3)]. This control level cannot be less stringent than any applicable new-source performance standard or emission standard for hazardous pollutants, but it can be more stringent. Because this determination is made on a case-by-case basis, it is possible to determine the economic consequences of various proposed emission limitations. However, such an analysis may not be performed for a variety of reasons including lack of resources available to the control agency.

States that can demonstrate attainment of national ambient air-quality standards can use any mix of control measures they choose and are free to use the most cost-effective means possible. For states designated nonattainment that were required to submit implementation-plan revisions in 1979, the act specifies that the plan contain identification and analysis of its economic, energy, and social effects and a summary of public comment on the analysis [§172(a)(9)]. The state plans examined by the commission varied in their response to this requirement.[40] Virtually all plans discussed the costs of the measures; however, few provided great detail on the employment and energy consequences of the proposed plan.[41]

As opposed to the standards for gasoline automobiles, the EPA may consider cost in establishing particulate-emission standards for diesel passenger vehicles and emission standards for all pollutants from trucks. For nitrogen-oxides emissions from diesel passenger cars, the EPA may grant a waiver of up to four years. One factor to be considered in its decision is the cost of achieving the required emission limitation. Once the EPA has established truck-emission standards, a manufacturer that cannot meet standards can continue to build vehicles that exceed the standards if a noncompliance penalty is paid.

In addition to consideration of economic effects required by the act, the EPA routinely has required economic analyses of all major regulatory actions since 1974. The economic analyses were requested by EPA administrators to ensure that economic factors were taken into consideration to the extent allowed by the act in establishing proposed and final rules.

The EPA's policy of requiring economic analyses of proposals and rules served as the model for Executive Order 12044, issued by the president in March 1978, requiring other government agencies to adopt similar procedures. Among the requirements of the order is the preparation of a regulatory analysis when a regulation is expected to have major economic consequences for the general economy, individual industries, geographical regions, or levels of government.

The criteria established by the EPA for identifying those proposed regulations that have the potential for major economic effects, and that should therefore be subject to regulatory analysis, include the following:

Additional costs of compliance, including capital investments, totaling $100 million within any one of the first five years of implementation or, if applicable, within any calendar year up to the date by which the law requires attainment of the relevant pollution standard;

Total additional cost of production of any major industry product or service exceeds 5 percent of the selling price of the product;

Requests by the administrator for such an analysis.

On 17 February 1981, Executive Order 12044 was replaced with a new executive order that:

Imposes, to the extent permitted by law, a requirement that agencies choose regulatory goals, set priorities to maximize benefits to society, and choose the most cost-efficient means among legally available options for securing these regulatory goals;

Requires all agencies to prepare, for each major rule, a regulatory-impact analysis that will be designed to permit an accurate assessment of the potential costs and benefits of each major regulatory proposal, including alternatives;

Authorizes the director of the OMB to review proposed and final agency regulations and regulatory-impact analyses for consistency with the order.

Economic-Incentive Approaches to Air-Pollution Control

The principal method used to control emissions under the act is direct regulation. Under this method, a source is directed by the control agency to meet specific emission limitations. Alternative approaches based upon economic incentives also have been discussed. The basic premise underlying economic-incentive approaches is that market forces—such as profit incentives or avoidance of fees—can be used to reduce pollution to desired levels more efficiently than direct regulation.

Prior to 1976, economic-incentive approaches to air-pollution control for stationary sources were discussed primarily in academic papers and journals. Few such proposals were considered by control agencies. Since 1977, however, not only have such approaches begun to be seriously considered by control agencies and big industry, but several have been adopted for use in meeting the act's requirements.

The use of these approaches in air-pollution-control programs has just begun. This section describes the approaches currently being used or considered and the experience with each.

The Bubble Policy

The bubble policy, which the EPA adopted in December of 1979 as an alternative means for existing facilities to comply with state implementation-plan requirements, is perhaps the best known economic-incentive approach to air-pollution control. The traditional regulatory system requires a facility to ensure that each unit emitting air pollutants complies with emissions standards; the bubble policy treats the facility as if all emissions came from a single stack—that is, as if a bubble covered the entire facility.

Thus, a facility that may emit pollutants at several points is able to control emissions to a greater-than-required level at one point and to control emissions to a lesser-than-required level at another so that the overall emission-reduction requirement of the state implementation plan is met. This flexibility may permit the facility to use a potentially more-cost-effective mix of emission-reduction measures.

The EPA adopted this policy as a result of studies suggesting that substantial cost savings could be achieved. One study, completed by Maloney and Yandle for fifty-two Du Pont Company plants, showed that allowing the use of the bubble concept at each of the plants to achieve 85 percent emission reductions would reduce costs of emission control by 35 percent over point-source controls.[42]

Implementation of the bubble policy by the EPA has allowed:

Trades among different categories of sources of volatile organic compounds within a plant;[43]

Firms to comply with existing state emission limits for individual operations by averaging emissions on a daily basis so long as the firm does not exceed total allowable plantwide emissions;[44]

Multiplant bubbles.[45]

According to the EPA, in all of these actions, resulting emissions will be equal to the emissions that would have occurred if specific point-source reductions had been required; however, substantial savings in costs should be realized. The agency stated that Du Pont's proposal to control five major volatile organic-compound emission points to more than 97 percent efficiency in exchange for relaxed controls on more than 200 sources of fugitive emissions that are difficult to control would probably be approved, saving $12 million in capital costs. More than 100 chemical firms in New Jersey are expected to utilize the state's bubble policy within six months of its approval by the EPA, which is expected by March 1981.[46]

The action on can-coating operations is estimated by the EPA to result in savings of $107 million in capital expenditures, $208 million per year in operating costs, and 4 trillion Btu's of natural gas per year because certain types of add-on pollution-control equipment will no longer be needed.[47]

The Narragansett Electric Company multiplant bubble permits the use of high-sulfur oil (2.2 percent sulfur) at one plant when the second plant is burning natural gas or is not operating, instead of 1 percent sulfur oil being used at both plants. The action will result in a savings of $3 million, reduce the use of imported oil by 600,000 barrels per year, and reduce sulfur-dioxide emissions by 30 percent.[48]

According to the EPA, as of 1 February 1981, there were sixty-nine bubbles in the EPA's reporting system, of which twenty-five bubble applications had been formally submitted, forty-three were under development, and one had been approved.[49] Two of these applications, in addition to the Narragansett bubble, involve multiplant proposals.

The EPA bubble policy contains some restrictions. The application for the bubble must receive approval through the state implementation-plan revision process, which can require additional time to obtain approval. The source must demonstrate that the proposed bubble trade will equal current regulations—in terms of both air quality and enforceability. Trades can be made only between emissions of the same pollutant; a firm cannot emit more of a hazardous pollutant in return for an equal reduction of a less-dangerous one. The source (or sources) must comply with applicable local, state, and federal air-pollution-control requirements or be on approved compliance schedules for all sources involved in a proposed bubble. Also, the EPA has not allowed new sources to use the bubble policy to relax control requirements under new-source-performance standards in return for greater control of emissions from existing sources.

The EPA's most recent actions—in particular, approval of the New Jersey state implementation plan—have attempted to accelerate the approval process for applications requesting the bubble approach. The EPA recently extended the bubble policy to areas that are not in attainment of the national ambient air-quality standards and that do not have an approved plan for attainment. To reduce the delay in processing the application, the

EPA has reduced the modeling requirement.[50] However, other issues remain that may hinder the use of the bubble concept, concluding the level of resources needed by industry and government to review and approve applications and the utility of the policy for volatile organic compounds if more-stringent controls are imposed on existing sources as a result of the 1982 implementation-plan revision requirement.

Emission Banking

Emission-reduction banking complements the offset and bubble policies. The EPA initiated this policy in 1980 in an attempt to encourage greater economic efficiency in meeting the requirements of the Clean Air Act.

Banking is an accounting and administrative process by which a firm can receive credit for reducing its emissions by amounts greater than required in the state implementation plan. The resulting credit forms the basis of the banking program. The firm that deposits a credit in the bank can use the credit to make a profit by selling it to another firm that needs an offset, or it can use the credit itself—as an offset, in a bubble application, or in some instances to satisfy PSD requirements.

The banking program is intended to assist in reducing the uncertainty and transaction costs that confront a new facility that is attempting to find sufficient offsets to locate in a nonattainment area. Offsets may be difficult to obtain in a limited number of cases, particularly for a source locating a new facility in an area where it has no existing facilities. An emission bank could assist sources in locating offsets in these situations. A successfully operated bank could help spur economic development in an area and could provide a competitive advantage over areas without banking programs.

A banking program also could, in theory, facilitate the use of the bubble policy to help existing firms to minimize the cost of complying with emission limitations. The existence of credits in a bank could allow a multiplant bubble to be created or a firm to create a bubble in the future.

Several communities have begun to establish emission banks. In Louisville, Kentucky, the bank established in June 1979 has had twenty-two deposits and six withdrawals. The San Francisco Bay–area bank established in January 1980 has had only one deposit and no withdrawals. The Puget Sound bank has had no formal deposits. In the Houston area, a private organization is in the process of developing a banking program. The Chamber of Commerce in Oklahoma City helped find companies willing to create offsets and brought the companies together with General Motors, which wanted to build an assembly plant in the city. The EPA has provided grants to several cities to develop clearinghouses to support trading markets in areas that have not attained national standards.[51]

One potential difficulty for emission banking is the reluctance of firms to notify the banking agency of the existence of credits. A firm may not want a competitor to expand with its emission credits, or it may not want to divulge the existence of an inherently low-polluting production process.

Other potential impediments to emission banking have not yet been resolved. One is the length of time a source could keep an emission-reduction credit in the bank for later use. Some proposals have set time limits within which a source must use its credits before they would become available to other sources in the area. Another is ensuring that the reductions placed in the bank are permanent and enforceable; if they are not, the utility of the program would be undercut.

Transferable Emission-Reduction Assessments

As part of its investigation of economic incentives, the commission studied the concept of transferable emission-reduction assessments (TERA). These assessments would be used primarily in areas that could not attain the ambient air-quality standards with reasonably available control technology and are a combination of the present regulatory system and the bubble concept. Under this approach, each polluting source would be required to install reasonably available control technology. Each source in the area would then be assessed a portion of the remaining reduction necessary to achieve the ambient standards. The additional emission reductions required could be obtained in three ways: (1) A source could reduce its own emissions; (2) purchase the emission reduction from another source, directly or through a third party or broker; or (3) pay an emission fee to the state or local air-pollution-control agency, which would use the fee to obtain reductions from other sources.

Two of the commission's regional studies evaluated the TERA concept. Three areas in the Ohio River Valley could not attain the annual total suspended-particulates ambient air-quality standard by applying reasonably available control technology. Under TERA, two of the three areas reduced costs by $4,190 and $89,171 from levels of $112,805 and $350,299, respectively. In another area, this approach made no difference in the cost of achieving the standard.[52] In the South Coast Air Basin, the commission study applied the TERA approach to hydrocarbons and nitrogen oxides to examine its effectiveness in reducing ozone levels. The study concluded that TERA would not work well in this region because the technology needed for further emission reductions was available for so few sources that significant trading could not occur.[53] Clearly, the commission studies are not conclusive regarding the merits of TERA. Analyses by the EPA have shown that TERA offers significant cost savings over a direct regulatory approach.[54]

The disadvantages of this approach include increased data collection and record keeping. In addition, the commission studies indicated that accounting for the required reductions was a potential source of controversy between industry and government.

Emission Fees

Certain economists and academic commentators have contended that an effective way to reduce pollution would be to impose a fee on a source rather than to regulate it directly. The fee would be set sufficiently high to make emission control more economical than payment of the fee, thereby resulting in achievement of the desired air-quality standard.

A fee program could be based on the level of emissions, the type of industrial process, or on the fuel used in an industrial process. The fee also could be applied in various ways. For example, the fee structure could distinguish between new firms and existing firms. The fee could be applied to specific sources or to an entire industrial category; it also could be charged on a geographic or a seasonal basis.

While emission fees have been studied extensively, they have rarely been used. The Navajo Indian Tribal Council attempted to impose a fee on sulfur-dioxide emissions. However, the program has never been implemented because of legal challenges and lack of approval by the DOI.

At the present time, the City of Philadelphia is proposing to impose emission fees and to distribute the revenue to certain sources in the area. The fee proposal is intended to encourage firms to reduce emissions on their own without government intervention. Under this proposal, companies with high marginal costs of control will be charged a fee for each ton of emissions released, while companies with low marginal control costs will be paid a subsidy for each ton of emissions reduced. Such a system minimizes the need for the government to make decisions about ways to allocate revenues. However, the government must determine which firms are subject to the fee and which firms receive the subsidy. Once this decision is made, the government would collect fees and distribute revenues. Philadelphia currently is studying an adjustment mechanism for the fee that is tied to ambient levels monitored. This could help ensure establishment of fair and appropriate fees and equitable distribution of the proceeds. The system initially would be used only on sulfur-dioxide emissions and may be ready for implementation by 1985.

Several air-pollution-control agencies impose fees as revenue-raising measures rather than as substitutes for regulation. Such fees usually are small and have no effect on pollution reduction. Several foreign countries including Japan, the Netherlands, West Germany, and Norway have established experimental fee systems to raise revenues and, in some instances, to reduce emissions. In 1979, East Germany established emission fees for 113 different air pollutants.[55]

Problems associated with emission fees include the difficulty of establishing appropriate fees and ensuring that accurate monitoring data are collected. If the fee is too low, appropriate air-quality goals will not be achieved. If it is too high, overcontrol and excess cost will result. If adequate monitoring data are not available, the fee payment could be either lower or higher than necessary to achieve air-quality goals. Another problem is the disposition of the revenue collected through the imposition of the fee. As the Philadelphia proposal suggests, the fee could be used for subsidies to control emissions. It could also be used to finance the operation of control agencies or as general revenue for the government.

The commission's regional studies simulated the effect of imposition of emission fees on stationary sources and compared this approach with direct regulation. These studies were undertaken for control of sulfur dioxide in the Four Corners region; for sulfur dioxide, nitrogen oxides, and hydrocarbons in the Los Angeles region; and for hydrocarbons in the Twin Cities–St. Cloud region. The Four Corners study included a limitation that the Class II PSD increments should not be exceeded. The Twin Cities-St. Cloud study included the limitation that the emissions under the emission-fees scheme could not exceed the emissions allowed under the direct regulations in the state implementation plans. The fee was assumed to be the average cost of control of a plant in the Four Corners region, and it was assumed to be the cost of the level of maximum control (or the highest individual source's cost of control) for the other two regions.

The Four Corners regional study estimated the fees at $0.90/ton, $1.00/ton, and $1.20/ton for sulfur dioxide at the air shed, the state level, and the regional level, respectively. These fees translate to about $2,000 per ton of sulfur dioxide emitted, which is high compared to per-ton costs for sulfur-dioxide removal in other areas. Nonetheless, cost comparisons showed that emission fees for the three geographic areas would cost $90 million, $180 million, and $359 million, respectively, while the cost of best-available-control technology and best-available-retrofit technology for the entire region would be $320 million, and the cost of uniform emission reductions of 96 percent from uncontrolled levels would be $670 million. The sulfur-dioxide emissions under a fee system would be substantially higher at the air shed and the state-level fees and somewhat lower at the regional-level fees compared to emissions under the existing requirements.

The Los Angeles study found that the cost of emission fees would be greater than the costs of direct regulation because the fee was established at the highest marginal cost of control for all three pollutants. An emission fee on hydrocarbons could lead to annual revenues of approximately $11 billion and total capital costs of $3.7 billion, which is more than the capital cost under the direct-regulation approach. Because of the level of the fee imposed, emission reductions were greater under the fee approach than under direct regulation.

The Twin Cities–St. Cloud study found that emission fees led to a 25 percent reduction in control costs compared to costs for direct regulation requiring application of reasonably available control technology. However, the annual fee payments would exceed the capital expenditures by a factor of ten. The levels of emissions under the fee approach were the same as those under direct regulation.

Subsidies

Unlike the federal-grant program for construction of municipal waste-water-treatment facilities, the act does not provide subsidies for capital expenditures needed to comply with the act's requirements. Proposals for direct subsidies to sources have not been made. However, indirect subsidies are available under the Internal Revenue Code and under many state laws. Two federal tax programs exist to reduce the pollution-control costs to firms: rapid amortization and the investment-tax credit.

Section 169 of the Internal Revenue Code provides that eligible pollution-control equipment can be amortized over a five-year period even though the actual useful life or normal depreciation period is longer. The present tax advantages of rapid amortization resulted from the 1978 amendments to the Internal Revenue Code. Prior to 1978, firms had not used this provision to any significant extent because it was mutually exclusive with the provisions of the investment-tax credit (a 10-percent credit against tax obligation for investments in plants and equipment). After the 1978 amendments allowed both rapid amortization and investment-tax credits, the use of rapid amortization increased significantly.[56]

Industrial development bonds are long-term bonds issued by a state or municipality or an industrial-development authority for projects with a public purpose. When the proceeds of the bonds are used in certain exempt activities—including air-pollution-control facilities—the interest on the bonds is tax-exempt to the buyer. The difference between exempt and nonexempt bonds can lead to a 3–4 percent difference in interest rate.[57] The proceeds of the bond issue are used to build the firm's pollution-control system, and all payments of interest and principal to bondholders are provided by the same firm. In effect, no municipal funds are involved in the transaction. All but two states (Hawaii and Washington) have authority to issue tax-exempt bonds for pollution-control investments.

Prognosis for the 1980s

A little more than a decade ago, Congress passed the Clean Air Act, the first major substantive law designed to preserve the environment. Since that time, the absolute level of improvement for the most widespread air pollutants has been significant. Between 1974 and 1978 there was an 18 percent reduction in the number of days during which air quality in twenty-three major metropolitan areas was classified as unhealthful. Nationwide, between 1973 and 1978, average annual concentrations of carbon monoxide decreased by 33 percent, sulfur dioxide by 20 percent, and suspended particulates by 7 percent.

More significant than the level of absolute reductions, however, is the difference between pollution levels and those that would have occurred if major control efforts had not been required during the 1970s. While it is impossible to state precisely what pollution levels would be if the act had not been passed, it is clear that for a number of pollutants the level of emissions would now be several times as great in many areas.

The prognosis for the 1980s reflects a need for the control of air pollution to be a continued national priority. During the 1980s, millions of people will continue to live in areas exceeding healthful levels for ozone and suspended particulates. In at least one major urban area—Los Angeles—the ozone problem appears to be intractable and no reasonable level of effort is likely to result in attainment of the standard in that area.

The need for continued substantial investments comes at a time when all Americans are concerned about economic problems of a magnitude experienced only a few other times in the history of the nation. Thus, it is incumbent upon policymakers to ensure not only that air quality is good but also that an appropriate balance exists among all of the nation's priorities.

Notes

1. B. Ferris and F. Speizer, "Suggested Criteria for Establishing Standards for Air Pollutants" (Report prepared for the Business Roundtable Air Quality Project, Washington, D.C., 1980).

2. U.S. Congress, House, Committee on Interstate and Foreign Commerce, *Clean Air Act Amendments of 1977: Report to Accompany H.R. 6161.* 95th Cong., 1st sess., H.R. 95–294, 1977.

3. SRI International, "Report on Health-Effects Research and Standard Setting at EPA" (Report to the National Commission on air Quality, Washington, D.C., Contract No. 25a–AQ–8044, 1980).

4. Ibid.

5. Ibid.

6. Ibid.

7. Ibid.

8. U.S. Congress, House, *Clean Air Act Amendments.*

9. SRI International, "Report on Health-Effects Research."

10. E.F. Denison, *Effects of Selected Changes in the Institutional and Human Environment upon Output per Unit of Input* (Washington, D.C.: The Brookings Institution, 1978).

11. E.F. Denison, *Accounting for Slower Economic Growth: The United States in the 1970s* (Washington, D.C.: The Brookings Institution, 1979).

12. SRI International, "Report on Health-Effects Research."

13. Council on Environmental Quality, *Environmental Quality—1979* (Washington, D.C.: Government Printing Office, 1979).

14. Ibid.; and G.L. Rutledge and S.L. Trevathan, "Pollution Abatement and Control Expenditures, 1972–1978," *Survey of Current Business* 60 (1980):27.

15. G.L. Rutledge and B.D. O'Connor, "Capital Expenditures by Business for Pollution Abatement, 1977, 1978, and Planned 1979," *Survey of Current Business* 59 (1979):20; and Council on Environmental Quality, *Environmental Quality.*

16. Rutledge and O'Connor, "Capital Expenditures."

17. Ibid.; and McGraw-Hill Publishing Company, *Thirteen Annual McGraw-Hill Survey of Pollution Control Expenditures* (New York, 1980).

18. McGraw-Hill, ibid.

19. Rutledge and O'Connor, "Capital Expenditures."

20. A.M. Freeman, "The Benefits of Air and Water Pollution Control: A Review and Synthesis of Recent Estimates (Report to the Council on Environmental Quality, Washington, D.C., 1979).

21. See L. Lave and E.P. Seskin, *Air Pollution and Human Health* (Baltimore: Johns Hopkins University Press, 1977); T.E. Waddell, *The Economic Damages of Air Pollution* (Washington, D.C.: Environmental Protection Agency, 1974); B.C. Lui and E.S. Yu, *Physical and Economic Damage Functions for Air Pollutants by Receptor* (Corvallis, Or.: Environmental Protection Agency, 1976); K.A. Small, "Estimating Air Pollution Costs of Transport Models," *Journal of Transportation Economics* 2 (1977):111; Crocker, Schultze, Shaul, and Kneese, *Methods Development in Assessing Air Pollution Control Benefits, Volume 1: Experiments in the Economics of Epidemiology* (Washington, D.C.: Environmental Protection Agency, 1979); and H.T. Heintz, S. Hershaft, and G.C. Horak, "National Damages of Air and Water Pollution" (Report for the Environmental Protection Agency, Washington, D.C., 1976).

23. Crocker, Schultze, Shaul, and Kneese, *Methods Development.*

24. Ibid.

25. Freeman, "Benefits of Air and Water Pollution Control."

26. SRI International, "An Estimate of the Nonhealth Benefits of Meeting the Secondary National Ambient Air Quality Standards" (Report to the National Commission on Air Quality, Washington, D.C., Contract No. 27a–AQ–8060, 1981.

27. Ibid.

28. Ibid.

29. Ibid.

30. Ibid.

31. Ibid.

32. Ibid.

33. Randall, Ives, and Eastman, "Bidding Games for the Evaluation of Aesthetic Environmental Improvement," *Journal of Environmental Economics and Management* 230 (1974).

34. Brookshire, d'Arge, Shultze, and Thayer, *Methods Development for Assessing Tradeoffs in Environmental Management, Volume II: Experiments in Valuing Nonmarket Goods* (Washington, D.C.: Environmental Protection Agency, 1979).

35. National Economic Research Associates, *The Buisness Roundtable Air Quality Project: Cost Effectiveness and Cost-Benefit Analysis of Air Quality Regulations* (New York: National Economic Research Associates, 1980).

36. R.A. Leone and D.A. Garvin, "Regulatory Cost Analysis: An Overview" (Report to the National Commission on Air Quality, Washington D.C. 1981).

37. Ibid.

38. Benefit Methodology Evaluation Panel, "Report of the Benefit Methodology Evaluation Panel to the National Commission on Air Quality" (Washington, D.C., 1980); and Sterling Hobe Corp., "Survey of Methods Measuring the Economic Costs of Morbidity Associated with Air Pollution" (Report to the National Commission on Air Quality, Washington, D.C, Contract No. 27a–AQ–8060, 1980).

39. Dames and Moore, "An Investigation of Prevention of Significant Deterioration (PSD) of Air Quality and Emission Offset Permitting Processes (A Study of PSD and Offset Permits)" (Report to the National Commission on Air Quality, Washington, D.C., Contract No. 1a–AQ–7133, 1981).

40. Pacific Environmental Services, Inc., "An Overview of the SIP Review Process at the State Level and the SIPs for Particulate Matter, Sulfur Dioxide, and Ozone: Study of the 1979 State Implementation Plan Submittals" (Report to the National Commission on Air Quality, Washington, D.C., Contract No. 3a–AQ–7353, 1980).

41. Urban Systems Research and Engineering, Inc., "An Institutional Assessment of the Implementation and Enforcement of the Clean Air Act:

Buffalo Case Study" (Report to the National Commission on Air Quality, Washington, D.C., Contract No. 19g–AQ–9130, 1980).

42. M.T. Maloney and B. Yandle, "The Estimated Cost of Air Pollution Control under Various Regulatory Approaches," unpublished paper. Clemson University, Department of Economics, Clemson, S.C. 1979.

43. Environmental Protection Agency, "Air Pollution Control: Recommendations for Alternative Emission Reduction Options within State Implementation Plans: Proposed Revision to the New Jersey State Implementation Plan," *Federal Register* 45 (24 November 1980):228.

44. Environmental Protection Agency, "Notice of Policy Memorandum: Compliance with VOC Emission Limitations for Can Coating Operations," *Federal Register* 45 (8 December 1980):237.

45. Environmental Protection Agency, "Proposed Rulemaking: Revision to the Rhode Island State Implementation Plan," *Federal Register* 45 (1980):208.

46. Environmental Protection Agency, *The Bubble Clearinghouse* (Washington, D.C.: Environmental Protection Agency, 1981).

47. Environmental Protection Agency, *The Bubble Clearinghouse* (Washington, D.C.: Environmental Protection Agency, 1980).

48. Environmental Protection Agency, "Air Pollution Control."

49. Environmental Protection Agency, *Bubble Clearinghouse.*

50. Environmental Protection Agency, *Environmental News* (16 January 1981).

51. Environmental Protection Agency, *Draft Regulatory Analysis of the Heavy Duty Diesel Particulate Regulations* (23 December 1980).

52. PEDCO Environmental, Inc., "Analysis of Emission Reductions and Air Quality Changes for Alternative Development Scenarios: Ohio River Valley Regional Study" (Report to the National Commission on Air Quality, Washington, D.C., Contract No. 110–AQ–7695, 1980).

53. Environmental Protection Agency, *Bubble Clearinghouse.*

54. D. Foster, Environmental Protection Agency, memorandum to M. Low, National Commission on Air Quality (14 November 1979).

55. Los Alamos Scientific Laboratory, "Alternative Policy Evaluations: Four Corners Regional Study" (Report to the National Commission on Air Quality, Washington, D.C., Contract No. 14q–AQ–7721, 1980).

56. TACA Corporation, "The Economic Impact of Regulating Air Pollution in the Ohio Valley: Ohio River Valley Regional Study" (Report to the National Commission on Air Quality, Washington, D.C., Contract No. 11q–AQ–7695, 1980).

57. Radian Corporation, "Analysis of Innovation Policy Alternatives: Study of Air Pollution Control Technology" (Report to the National Commission on Air Quality, Washington, D.C., Contract No. 15d–AQ–7421, 1980.

10 Technology Assessment and Nuclear Energy: Moving Beyond Three Mile Island

Deborah D. Roberts

Three Mile Island and Technology Assessment

The events at Three Mile Island, Pennsylvania, beginning on 28 March 1979, were undeniably a turning point for civilian nuclear energy and perhaps for the governance of nuclear technology in the United States as well. TMI provided nuclear-energy opponents with a symbol far more potent than a Seabrook plant in construction or an albacore kill from the release of cooling waters at Diablo Canyon. However, nuclear-industry proponents contend that TMI's importance has been greatly exaggerated.[1] No one was physically injured, and the official Kemeny Presidential Commission concluded that there would be no detectable additional cases of cancer, developmental abnormalities, or genetic ill health from the accident.[2] Still, TMI diminished public confidence in nuclear technology and its regulation and, most important, prompted changes in assessments of nuclear power. This chapter examines the politics of assessing nuclear-power technology in the aftermath of TMI. TMI remains crucial because it marks fundamental changes in the politics of assessing and regulating nuclear technology.

Nuclear Technology Assessment Prior to TMI

In the first twenty years of commercialization, nuclear power was governed by a concentrated, centralized decision-making system. States deferred to the federal level, and Congress and the courts deferred to the executive branch. Public policy during the 1950s and 1960s promoted civilian nuclear technology at a time of global conflict and international science and technology competition. In his 1954 Atoms for Peace Address, President Dwight Eisenhower urged that nuclear power be commercialized as quickly

The author gratefully acknowledges her debt to James D. Carroll, professor at the Maxwell School of Citizenship and Public Affairs, Syracuse University, for his generous advice and invaluable assistance in the preparation of this chapter.

as possible and thereby promote a "new era of progress and peace."[3] The Atomic Energy Commission (AEC) had been established under the Atomic Energy Act of 1946. Congress subsequently passed the Atomic Energy Act of 1954 and increased the AEC's broad authority both to promote and regulate civilian nuclear power. Under the act, the unique and powerful Joint Atomic Energy Commission was a proponent and protector of AEC's privileged status within Congress. Indeed, through the 1970s, aggressive federal policy—a technology push—literally created a nuclear-power industry out of a reluctant and cautious electric-power industry. The federal government used inducements like R&D grants, sponsored nuclear-reactor demonstrations and prototypes, provided nuclear-fuel subsidies, and limited liability to reduce the commercial risks of nuclear energy. By 1968, half of all orders for new power plants were for nuclear reactors. Still, by 1970, the nuclear industry, supported for decades by the federal government, contributed only 1 percent of the total national energy produced.

Disagreement over entrusting both regulatory and promotional responsibility for nuclear energy to the AEC, however, led to the Energy Reorganization Act of 1974. Congress replaced the AEC with the Nuclear Regulatory Commission (NRC), which now has primary federal responsibility for regulating nuclear power, and the ERDA (now part of the DOE), which has the promotional and technology-development responsibilities.[4] The Joint Atomic Energy Commission was also dismantled, and congressional oversight and jurisdiction for nuclear power was given to several committees. Congress nevertheless continued to support the nuclear-power industry. For example, in 1975, Congress extended the Price-Anderson Act, limiting commercial liability for nuclear accidents. More important, the NRC, like the AEC previously, had broad administrative discretion. Congress never provided explicit statutory standards for policy issues such as nuclear-reactor siting and safety or managing nuclear wastes.

The courts affirmed the broad delegation of power, responsibility, and discretion to the executive branch to regulate nuclear power. In *Power Reactor Development Co.* v. *International Union of Electrical, Radio and Machine Workers,* the Supreme Court upheld the AEC's authority under the Atomic Energy Act to allow a two-step licensing process in which only provisional safety findings were required at the construction-permit stage.[5] Moreover, Justice William Brennan voiced his belief in the ability of technologists to solve problems of nuclear safety: "Problems which seem insufferable now may be solved tomorrow."[6] Lower federal courts adhered to the *Power Reactor* view that judges must "show extreme deference to AEC and NRC decision making beyond that accorded other agencies."[7] The Supreme Court did not have occasion to rule again on the regulation of nuclear power until challenges to the NRC's authority in 1977. In *Vermont Yankee Nuclear Power Corp.* v. *Natural Resources Defense Council, Inc.,* the Court again rebuffed judicial intervention in science-policy disputes over nuclear energy.[8] The Court ruled that the judiciary should not impose procedural requirements on agency rule making other than those congressionally mandated or developed by an agency itself. The Court subsequently upheld, in *Duke Power Co.* v. *Carolina Environmental Study Group, Inc.,* the constitutionality of the Price-Anderson Act of 1957, which imposed limits on liability for damages incidental to a nuclear accident.[9]

Congressional delegation of authority and judicial deference to the AEC and the NRC led to an institutional assumption of safe nuclear power, or at least a presumption that the benefits outweigh any unclear and therefore improbable risks of nuclear energy. The risks of nuclear technology were presumed to be manageable and commensurate with scientific and technological developments. For example, to encourage private-utility ownership, the AEC after 1954 shifted its safety policy from one based on remote siting to a defense-in-depth containment policy, thereby allowing nuclear-plant construction even close to densely populated urban areas. The NRC continued many of the AEC safety-risk-related policies. Analysis of the safety systems, for instance, continued to be delayed until the last possible stage of plant licensing on the assumption that any attendant technical problems could be resolved while construction proceeded. The NRC furthermore allowed over 200 problems existing in all reactors or all reactors of a specific design to be classified as generic problems, which in turn cannot be cited to deny an individual plant reactor license.[10] As a result, no power-plant reactor has ever been abandoned in the last licensing stage, and no operating license has ever been denied.

When nuclear power was first commercialized, the AEC fully recognized that a major accident was possible. A 1957 worst-case scenario, involving a 50 percent release of a nuclear plant's core materials into the atmosphere under the worst possible weather conditions, estimated the social cost only upon the need to evacuate 460,000 people in a 760-square-mile area.[11] The AEC and the NRC regarded engineered safeguards—such as safety equipment, the containment structure, and the ability to evacuate the immediate population zone (usually a three-mile radius)—as a permissible trade-off and justification for the siting of nuclear plants near heavily populated areas. However, no attempt was made to calculate the probability of a reactor accident. Indeed, only in 1975 was the first attempt made to assess both the probability and consequences of a major reactor accident.[12] Still, the *Reactor Safety Study,* commonly referred to as the Rasmussen study, or WASH-1400, was faulted for bias because most figures publicly released included only early fatalities and did not attend to the long-term health consequences of a nuclear accident.[13] An NRC review later cited the Rasmussen study's lack of generalizability, inadequate data base, and use of incorrect statistical methods.[14] Such initial nuclear-risk-assessment and safety programs, moreover, typically assumed that the physical design of nuclear plants mitigated against consideration of other possible sources of accidents such as terrorism, natural disasters, human error in both design and operations, and the deterioration of the physical plant's performance over time.

Nuclear technology assessment basically revolves around the question: What is the acceptable level of risk? Answering this question involves assessing and weighing the risks, designing and implementing measures to deal with those risks, and judging whether the risks are acceptable and the precautionary measures feasible. When nuclear energy was first commercialized, the third task was answered in the affirmative without full knowledge of the risks or the complexity of managing those risks. Perhaps that is not surprising because, as Harold Green points out, "Technology assessment, especially in the early stages of a technology, [often] involves a likely overweighting of benefits and underweighting of risks."[15] More recently, Joel Yellin concluded that "had a malevolent deity set out to test our system for overseeing technological decisions," he could not have come up with a better test than that of nuclear technology to show the "inherent weaknesses that prevent existing institutions from adequately controlling the risks" of developing technologies.[16]

The Political Significance of TMI

TMI undermined the public illusion of scientific consensus on nuclear-technology risk assessment. Early assessments in fact had not been subject to the full scrutiny of the scientific community: The Rasmussen study, for example, the authoritative work, was faulted for its lack of adequate peer review.[17] More crucial, TMI provided experiential data and demonstrated that risk calculus cannot be based solely on scientific judgments as to the likelihood of a natural event or technical malfunctions. Risk assessments depend on the integration of various scientific disciplines, many of which, such as the social sciences and human engineering, are neither as consensual nor have as well developed a theoretical base as nuclear physics. Moreover, pluralism and conflict exist within the scientific community itself. Thus the political choices and consensus underlying the regulation of technologies is exceedingly vexatious because of both disagreements on the scientific basis for regulation and the inexorably intense value conflicts attending regulation.[18]

TMI also raised the issue of how to translate risk assessments into routine industrial practices. The technical accident precipitating TMI had in fact occurred twice before, but that information was not disseminated to nuclear-reactor operators. Westinghouse had not been required to report to the NRC a similar incident that occurred at its reactor in Benznau, Switzerland. Likewise, the Davis Besse incident at Toledo Edison, which was analyzed by the utility company, Babcock and Wilcox, and the NRC, was not communicated to Metropolitan Edison at TMI.[19]

Overall, assessments of nuclear-technology risks appear to have lagged behind exploitation of the technology. The Rasmussen study, regardless of its limitations, was not even acted upon by the NRC or the nuclear industry. At the two nuclear plants the Rasmussen study analyzed, up to 90 percent of the risk of an accident was associated with the prospect of human error. As the NRC task force concluded, "transients, small loss-of-coolant accidents, and human errors are important contributors to overall risk, yet their study is not adequately reflected in the priorities of either the research or regulatory groups.[20] In contrast to the aggressive transfer of hardware advances, dissemination of management and assessment tools has not been as comparable.[21] One reason for what might be regarded as a low level of organizational learning is that the nuclear-power industry and the NRC remain preoccupied with nuclear-plant design and customization. In addition, within the public and private sector, responsibility for nuclear-safety management remains fragmented. The NRC's reporting system for abnormal operating events is itself an obstacle to assessing and minimizing the risks of accidents. Even with 2,000 to 3,000 reports each year, as the Rogovin study concluded, the NRC "does not distinguish the significant from the trivial."[22] Indeed, shortly before TMI, the GAO had faulted the NRC for its information overload and failure to identify important safety-related problems.

Two Different Models for Risk Assessment

Two main approaches have been taken to nuclear-technology risk assessment. The first, a prediction model, concerns the potential risks, costs, and benefits of nuclear energy. The second, a consequences model, focuses on the kinds of nuclear accidents and strategies of response, regardless of the probabilities of those accidents' actually occurring.

Technology assessment has normally concentrated on predicting the risks of nuclear accidents. The prediction model was implicit in and typified by the Rasmussen study, which centered on estimating the probability of nuclear accidents. The risks of nuclear technology were assumed to be equal to a nuclear accident's consequences multiplied by the probability of the occurrence of the accident. Accordingly, a nuclear accident that had catastrophic consequences might be deemed a low risk if the probability of the accident's occurring remained extremely remote.[23]

Risk assessment under the prediction model yields criteria for what regulatory actions ought to be taken to make those risks manageable and cost-effective. Once a risk has been quantified, then decisions can be made as to whether or not the risk is significant enough to warrant affirmative pre-cautionary actions—for example, whether the physical system can be de-signed with redundancy to offset its effects. The NRC uses such a design basis to analyze the safety of the engineering and siting of nuclear power plants. Class Nine accidents, for instance, include a combination of failures that lead to either a core melt down or severe containment failures. Their probability of occurrence is reduced to an expected cum acceptable low level of risk through plant design and safety features. The social consequences of a Class Nine accident and how they would be managed are nonetheless not considered in the licensing process.[24]

In contrast, the consequences model focuses on the socioeconomic and environmental impact of nuclear accidents with only minimal concern for the probabilities of those accidents' actually occurring. The consequences model was advanced by the 1978 NRC/EPA task force. It concluded that "there was no specific accident sequence that could be isolated as the one for which to plan, because each accident could have different consequences, both in nature and degree."[25] The consequences model thus focuses on responses to various possible nuclear accidents. Fundamentally, the model assumes that, since the consequences of a nuclear accident would be so great, institutions must be ready to respond to any event. Once a radiation release is a threat or a reality, how it happened (its cause) or its predicted likelihood (its probability) becomes only marginally important. In other words, the consequences model acknowledges that prediction of nuclear accidents is an imperfect science. While probabilities can be assigned, their actual incidences are not predictable in the same way that technologies—such as fossil fuels—have predictable and recurring effects in terms of annual mortality- and morbidity-rate increases. The consequences model therefore focuses on responses to possible, if improbable, accidents and thereupon designs and prescribes test drills, contingency procedures, and contacts between different governmental agencies and private organizations for responding to nuclear-industry emergencies.

Less than one year after the NRC/EPA task-force report, the problems with present emergency-management plans were graphically illustrated at the TMI. Confusion reigned. TMI demonstrated that even an event with no deaths or widespread radiation could be a traumatic accident. TMI also revealed how smaller initiating accidents, when compounded by human errors, could come very close to a catastrophe.[26]

TMI ostensibly eroded public trust in safe nuclear energy. Nuclear tech-nology assessment accordingly requires reestablishing public confidence in the capabilities of public and private organizations to predict nuclear risks and to respond effectively and efficiently to nuclear accidents. Several of the official studies stemming from TMI have thus stressed that an emergency-management system must be considered independent of plant-design factors in order to assure public health and safety.[27]

Emergency management, especially for the offsite consequences of a nuclear accident, has been historically neglected in the assessment and regulation of nuclear plants. The Federal Emergency Management Agency (FEMA), which is responsible for coordinating emergency responses to nuclear accidents, attributes the low priority of emergency management to "long-seated deficiencies of general emergency planning and preparedness programs at the federal, state and local government levels in the United States."[28] Certainly, emergency management is an extremely difficult and vexing task. What constitutes an emergency is ambiguous, and therefore the institutional roles and powers remain unclear. The extent of presidential authority, for example, in a case like TMI, depends on the incumbent's interpretation of implied and explicit presidential powers.[29] Also unclear is what the NRC's full powers are vis-a-vis that of a utility company. This ambiguity in institutional responsibilities results from fragmentation of governmental authority.

Emergency-management systems have therefore been proposed in part to clarify and structure public- and private-sector responsibilities and responses to potential nuclear accidents. In addition, emergency-management planning promises an opportunity for citizen participation in the regulation of nuclear power. Public participation in the regulation of nuclear power seems especially compelling since the attendant risks are localized: 3,336,000 Americans live within ten miles of an existing or planned nuclear plant.[30] Yet, as a GAO report showed, citizens living around nuclear power plants remain largely ignorant of the hazards and what they should do in case of a nuclear accident.[31] The GAO report further found that nuclear-facility operators have discouraged efforts to inform the public about the prospects and consequensces of a nuclear accident. With a touch of ironic coincidence, the GAO report—released two days after the advent of the TMI crisis—concluded: "There is only limited assurance that persons living or working near these nuclear facilities would be adequately protected in case of a serious, although unlikely nuclear accident."[32]

Finally, emergency-management planning would potentially democratize regulatory decision making over nuclear technology. Emergency-management planning requires an inherently intergovernmental policy, one which is vexingly difficult to attain because it depends on many autonomous public agencies—most with missions and skills unrelated to nuclear power.[33] TMI nonetheless epitomized for states and local communities that they are not bystanders in the federal and private-sector nuclear-power system. When accidents happen, state and local governments must respond.

The Intergovernmental Framework for Regulating
Nuclear Technology

Legislative, administrative, and judicial roles and responsibilities in regulating nuclear technology have historically developed in an incremental, fragmented, and at times, conflicting fashion. The intergovernmental network upon which emergency-management planning depends covers an array of relationships at the federal, state, and local levels:

Federal level:
Between federal agencies,
Within an agency or department (for example, divisions of the NRC and DHHS,
Between headquarters, regions, and field offices;

State level:
Between state and federal agencies,
Between states,
Between state actors and agencies;

Local level:
Between local government and the state,
Between local jurisdictions,
Between local-government units and community organizations (for example, police, fire, hospitals, and transportation),
With local private-utility company.

After TMI, this complex set of intergovernmental relationships appears to pose serious problems for coordinating governmental and private responses to nuclear accidents.

At the federal level, plans for assistance and coordination during peacetime, civilian power accidents have long existed. Yet, it remains unclear how those plans would translate into action at a time of an emergency and how to ensure that federal agencies neither shirk critical responsibilities nor intervene without authority. For example, the Interagency Radiological Assistance Plan (IRAP), first developed in 1961 and revised in April 1975, is a memo of understanding among thirteen federal agencies. At the time of TMI, some member agencies nevertheless did not know if IRAP was still in effect. Others were apparently even unaware of its existence, and no agency appeared certain as to who should initiate the plan. The NRC has primary responsibility for responding to emergencies. Other agencies—including the EPA Defense Civil Preparedness Agency, Federal Disaster Assistance Administration, Federal Preparedness Agency (now part of FEMA), and the Departments of Energy, Transportation, and Health and Human Services, as well as the FDA and the Public Health Service—also share some responsibility for planning and responding to nuclear accidents. In 1976, the Federal Preparedness Agency in the General Services Administration issued a federal response plan for peacetime emergencies and thereby sought to clarify federal responsibilities and to relate those responsibilities to the severity of nuclear accidents. For example, a Category 2 incident, posing a potential or widespread radioactive contamination, gives major operational responsibility to the NRC, the DOE, the Defense Department, and the Justice Department.[34] There were thus several master plans for federal coordination in effect at the time of TMI, yet all were subject to interpretation as to each agency's exact jurisdiction.

The National Academy of Public Administration's evaluation of the federal response plan described the inadequacies of present plans for federal responses to nuclear accidents: "Paper documents [are] . . . not backed up by political power, a strong constituency, budgetary power, technical capability, and history in the subject area."[35] Responsibilities for safety programs and emergency management are also fragmented within the NRC. The NRC has five equally powerful commissioners and five office directors with independent jurisdictions, which in turn leads to tensions within the NRC and, at times, paralysis of agency action. As the National Academy of Public Administration study concluded: "No individual or office in NRC is responsible for assessing the adequacy of the total emergency planning and response capability for each site."[36] Prior to TMI, the NRC had an executive management team to respond to nuclear emergencies, but it amounted to little more than a collection of representatives of each NRC office. In examining the NRC, the Rogovin study, like the National Academy of Public Administration study, thus emphasized the need for a single executive in command of emergency responses and for shifting management to the accident site as quickly as possible.[37]

Within the intergovernmental system, the primary responsibility for public health and safety resides with the states. States, however, have been largely reluctant to develop or require local communities to develop offsite emergency-management programs. To some extent, the NRC required private-utility operators to work with state and local governments as part of the licensing process.[38] Still, there have been few incentives for states to pursue a vigorous role in regulating nuclear power and preparing for the consequences of a nuclear accident. Furthermore, the judiciary has upheld the NRC's position that the Atomic Energy Act preempts the states from regulating radioactive emissions from nuclear power plants.[39]

The NRC has prompted some state action by establishing an elaborate concurrence process for state emergency plans. For example, in 1975, it established the Office of State Programs. In each of ten federal regions, a regional advisory committee for radiological-emergency response planning, at the invitation of a state, works with state officials to develop a plan for responding to nuclear accidents.[40] The committee recommends the state plan to the Office of State Programs and the Federal Interagency Central Coordinating Committee for approval. States must then conduct at least one exercise of every plan annually. What this concurrence process has yielded in substance remains debatable. As of October 1979, the NRC had approved sixteen state emergency plans. Yet, through its licensing process, the NRC had certified only four of seventy-one operating nuclear plants as conforming to its requirements for emergency-management plans.

Basically, the key to effective emergency response is adequate county and local, and not state, emergency plans. Governmental responsibilities at this level, however, are even more fragmented: 250 jurisdictions are covered by the FEMA's current ten-mile-radius emergency-planning zone. Several thousand local governments are within the fifty-mile-radius zone for long-term radiation effects. Moreover, prior to TMI, this intergovernmental planning system was overly centralized in the sense that local authorities were the last to be involved in emergency plans. The National Academy of Public Administration study accordingly recommended that "the community surrounding and including the site should be designated as the basic planning unit. The community should initiate all planning."[41] Obviously, such a grassroots approach to emergency management would be a drastic change in the regulation of nuclear power and would require more public participation, decentralization, and expenditure of local resources. It would, nonetheless, sensitize communities and local public agencies to the import of nuclear technology assessment and planning.

Institutional Changes since TMI

The presidential commission on TMI, as well as other official investigations, found that intergovernmental planning for radiological emergencies lacked both coordination and urgency.[42] Some progress has been made since TMI. On 5 October 1979, the NRC accepted in principle the recommendations of the NRC/EPA task force, which were based on a consequences model to emergency management and which entail establishing generic emergency-planning zones.[43] A ten-mile-radius emergency-planning zone is based on short-term plume exposure (external body exposure or inhalation). A radius of fifty miles is the standard for long-term ingestion exposure. These zones provide a basis for a management system but involve even more federal and local agencies in emergency planning.

Based on the Kemeny-study recommendations, in December 1979, President Jimmy Carter designated the FEMA as the lead agency for coordinating offsite radiological emergency preparedness.[44] However, the problem persists that, while the FEMA has the policy lead, the NRC has the expertise and the staff. Moreover, the FEMA has not been able to get requested appropriation levels and has had to rely heavily on personnel on loan from the NRC. The FEMA nevertheless did review state emergency-management plans in June 1980. The FEMA and NRC also created national evaluation criteria for utilities and state and local governments.[45]

The FEMA and NRC program has been extremely controversial. Its plans for mass evacuations have been criticized as too expensive and unwarranted on present estimates of risk. Industry representatives further contend that the new emphasis on emergency planning violates some of the recommendations of the Kemeny commission—in particular, that emergency plans be tailored to the degree of potential danger.[46] States and localities have also contested requirements for notification of 100 percent of the public inside a five-mile radius of the nuclear power plant within fifteen minutes of an emergency and for a near-site emergency-operation facility to be located within one mile of each power plant. The criticisms center not only on the disputed feasibility or advisability of such measures but on the economic costs of such plans for emergency management. The average implementation cost, as estimated by the FEMA, for state and local governments is about $1 million start-up per site. More densely populated sites could cost more than $2 million. Thus, concern arises not only over how to finance emergency-management plans but also over how resources will be allocated.[47] A dozen states have passed laws funding the plans through assessments on private nuclear facilities. Other options include requiring state and local emergency-response plans as a condition of licensing nuclear plants, thereby forcing the costs to be borne by the utility company and its customers; establishing a federal-grant program for state and local governments; and levying a federal tax on all utilities with nuclear power plants.[48]

The NRC also upgraded its emergency-planning regulations for the utility-company plant operators. An applicant or licensee must now submit emergency plans along with the state- and local-government plans for NRC approval. The utility operator must detail how emergency plans would be implemented and what drills would be conducted.[49] Moreover, an operating licensed reactor may be forced to shut down if deficiencies of offsite state and local plans are not corrected within four months. These changes, of course, have been extremely controversial with the nuclear industry. Instead of a one-time decision, nuclear-power licenses now depend on periodic renewal and continued commitments by the states and localities. The nuclear-power industry also has complained that such significant policy changes have resulted not from legislation or executive order but from memos of understanding and the NRC's view of its broad discretion.[50] Ultimately, the problem of clearly stated safety objectives rests with Congress. Still, Congress has broadly delegated authority to the NRC and dictated only that the

NRC regulate in a manner consistent with adequate protection for "public health and safety."[51]

Implementation of the new emergency-management policies was slow and difficult. Compliance with the NRC regulations for local warning systems was postponed from July 1981 until February 1982 because only seven plants had installed the sirens by the first deadline. Obstacles arose because of discontent by utilities, localities, and states. More fundamental, an effective emergency-management system presumes establishing clear governmental responsibilities, competent personnel, improved communication systems, commonly understood goals, and simple operating procedures.[52]

Post-TMI regulatory reforms have basically attempted to establish an effective emergency-management system on three sets of exercises or drills. Onsite drills are conducted by the NRC Operations Center in Bethesda and are now required before licensing any new plant. These drills are also done at older plants if pressure exists from states, localities, or operators to do so—as, for example, with plants in heavily populated areas such as the Tennessee Valley Authority's Brown's Ferry plant. The NRC drills include an outside contractor, the NRC Office of Inspections and Enforcement, representatives from other federal agencies with emergency radiological duties, and invited outside critics from the states. A second set of drills involves those held annually under state emergency plans. The third set of drills is also held annually between the utility operator and the localities in which states selectively participate.

The coordination problems of emergency management are illustrated by these three sets of drills. Most are neither smooth exercises nor should they be expected to be since the primary aim is a learning experience. Still, the basic institutional problem remains that no comprehensive, intergovernmental emergency-management system exists. Indeed, most legislative action in the wake of TMI has come at the state level.[53]

At the national level, the FEMA has completed the first phase for development of its master plan.[54] The NRC is now responsible for the technical coordination of any federal response. The DOE coordinates all offsite radiological monitoring, evaluation, and reporting activities of participating federal agencies. The NRC integrates the DOE evaluation with its own evaluation of the onsite situation into an overall assessment of the accident. The FEMA also coordinates all nontechnical aspects of the federal response and integrates state-developed plans. All state requests for resources from the federal government are channeled through the FEMA, and the FEMA coordinates the interagency public-affairs group controlling release of information to the public. During an emergency, the FEMA and the NRC are to exchange liaisons to prevent problems due to the NRC's focus on plant site and the FEMA's focus on state- and public-agency coordination. The FEMA master plan, however, has come at a cost to the FEMA's other operations since it agreed to work within the schedules and requirements of the NRC. The NRC required all operating plants to implement the new emergency-preparedness plans by 1 April 1981. Hence, "FEMA had to review, assess, and report to the NRC on more than 500 plans for some 50 nuclear power plant sites."[55] As of August 1981, the FEMA had completed less than half of the reviews.

Pluralistic Technology Assessment and the Prospect of Policy Paralysis

The fragmentation of nuclear-energy policy has led to a veritable policy paralysis that may prove intractable. Diverse interests, in contrast to the old concentrated technology-push alliance, may prove incapable of achieving agreement on regulating a now-established technology embedded in the economy. The New England states, the Southeast, and the Chicago area are all heavily committed to nuclear power. This policy paralysis may continue due to not only the polarization of interest groups but also to the sheer magnitude of the necessary institutional changes and factors such as inflation and declining energy demand. The nuclear-power industry was in serious trouble even before TMI. In 1973, forty-one nuclear power plants were ordered; in 1980, no new orders were placed. More important, since 1973 the net number of new reactor orders has remained at zero due to cancellations and postponements. In addition to the overall decline in electricity demands, nuclear-power-generating capacity has become extremely costly because of delays in licensing, huge construction-cost increases, a fourfold increase in uranium prices, and growing uncertainty over radioactive-waste-disposal policy and costs; the TMI cleanup itself is likely to cost over $1 billion, of which the utility's insurance covers only $300 million.[56]

Policy paralysis is evident in a number of areas. No agreement emerged among the numerous official studies of TMI as to what exactly ought to be done, even though all agreed that radical regulatory changes were needed. Perhaps the only consensus achieved was that determination of acceptable nuclear risks should not be the exclusive province of the NRC.[57] No fundamental changes have been made in the nuclear industry's structure or its relationship to the NRC. A number of industry reforms have occurred, such as the creation of the Institute for Nuclear Power Operations and the Nuclear Safety Analysis Center, which hold the potential for circulating information about minor technical accidents and safety problems among utilities. However, the impact of these organizational changes on the daily operations of the nuclear industry will take time to assess. Reforms such as those proposed for one-stage licensing, a formal office of public counsel and intervener funding, and a national operating company or consortium have all been rejected.

Furthermore, President Carter created a nuclear-safety oversight committee to review the NRC and industry approaches to nuclear safety, but it has not been continued under President Ronald Reagan's administration. Indeed, instead of President Carter's policy of nuclear power as a last resort, President Reagan has advocated the commercialization of breeder reactors and the expedited licensing of nuclear plants before the industry collapses. In addition, the NRC's budget has been cut 12 percent, with the burden falling on the safety technology staff and research even while new problems with conventional light-water reactors are being detected.[58]

Most crucial, few substantive organizational changes have been made in the formulation of nuclear-energy policy. No central executive authority or policy control has been established. In particular, the NRC has not been restructured. Criticisms that the NRC is "an organization that is not so much badly managed as it is not managed at all" persist.[59] Divisions among the commissioners and between the commission and agency staff remain. To be sure, some internal reorganizations and additions have taken place, such as the creation of new groups to review human factors and to monitor and evaluate the technical issues identified by TMI investigations. Nonetheless, the NRC continues to be preoccupied with promulgating and ensuring compliance with detailed regulations instead of achieving safety through performance specifications and development of achievable safety goals.[60] The NRC has also recently raised doubts over the adequacy of the International Atomic Energy Agency to prevent diversion of nuclear materials to atomic weapons under the Nuclear Non-Proliferation Act of 1978.[61] Finally, congressional oversight, which is critical to nuclear technology assessment and regulation, remains fragmented. Fragmentation of congressional oversight is not conducive to making hard choices in a policy area that no longer has a strong unified constituency.

In historical perspective, a certain irony is evident in the drift of nuclear-power policy. The move against the powerful alliance of the AEC and the Joint Atomic Committee broke up the nuclear-power oligarchy. Yet, by splitting the AEC into the NRC and the ERDA, and also by spreading congressional jurisdiction, greater access to the regulatory politics of governing nuclear technology was introduced. In theory, a pluralistic system is more responsible and responsive, but now the question is whether the policy arena is so differentiated and fragmented that it has become impossible to achieve consensus on any policy direction in the regulation of nuclear technology.

From a political perspective, TMI bode changes in nuclear-power technology assessment. The technical knowledge gained from TMI would appear to lead to more informed and conservative risk assessments. The public certainly appears to be less tolerant of nuclear-technology experts, and the scientific community has itself become more divided over the risks of nuclear technology. While incremental organizational changes have occurred, the challenge remains that of forging institutional innovations—creation of new institutions for conducting technology assessments and for developing and coordinating regulatory plans for managing nuclear technology.[62]

Notes

1. For the view of TMI as a phantom accident, see F.C. Olds, "Post TMI Plant Designs," *Power Engineering* (August 1980):54–62. For the exaggerated threat from radioactive release, see Andrew P. Hill,"Emergency Preparedness for What?" *Nuclear News* (April 1981):61–67.

2. John G. Kemeny et al., *The Presidential Commission on the Accident at Three Mile Island: The Need for Change, the Legacy of TMI* (Washington, D.C.: Government Printing Office, 1979).

3. President Dwight D. Eisenhower, "Special Message to the Congress Recommending Amendments to the Atomic Energy Act, February 17, 1954," in *Public Papers of the Presidents of the United States,* Dwight D. Eisenhower 1954, paragraph 38, p. 269 (Washington, D.C.: Office of the Federal Register, 1960).

4. 42 U.S.C. §§5801–5891.

5. *Power Reactor Development Co.* v. *International Union of Electrical, Radio and Machine Workers,* 367 U.S. 396 (1961).

6. Ibid.

7. Joel Yellin, "High Technology and the Courts: Nuclear Power and the Need for Institutional Reform," *Harvard Law Review* 94 (1981):515.

8. *Vermont Yankee Nuclear Power Corp.* v. *Natural Resources Defense Council, Inc.,* 435 U.S. 519 (1978).

9. *Duke Power Co.* v. *Carolina Environmental Study Group, Inc.,* 438 U.S. 59 (1978).

10. Statement by Robert Pollard in "Report of the Office of Chief Counsel on the Nuclear Regulatory Commission," in *Report of the President's Commission on the Accident at Three Mile Island* (Washington, D.C.: Government Printing Office, October 1979), p. 66.

11. U.S. Atomic Energy Commission, *Theoretical Possibilities and Consequences of Major Accidents in Large Nuclear Power Plants,* WASH-740 (Washington, D.C., 1957).

12. Norman C. Rasmussen et al., *Reactor Safety Study* (Washington, D.C.: Nuclear Regulatory Commission, 1975).

13. See, for example, Frank von Hippel, "Looking Back on the Ras-

mussen Report," *Bulletin of the Atomic Scientists* (February 1977):42–47; and Bruce L. Welch, "Deception on Nuclear Power Risks: A Call for Action," *Bulletin of the Atomic Scientists* (September 1980):50–54.

14. Harold W. Lewis, et al., *Risk Assessment Review Group Report to the Nuclear Regulatory Commission* (Washington, D.C.: Nuclear Regulatory Commission, 1978).

15. Harold P. Green, "Limitations on Implementation of Technology Assessment," *Atomic Energy Law Journal* 59 (1972):69.

16. Yellin, "High Technology and the Courts," p. 490.

17. See Lewis, *Risk Assessment.*

18. Dean Schooler, Jr., *Science, Scientists, and Public Policy* (New York: Free Press, 1971).

19. Michael Rogovin, et al., *Three Mile Island: A Report to the Commissioners and to the Public* (Washington, D.C.: Nuclear Regulatory Commission Special Inquiry Group, 1980), pp. 94–95.

20. Lewis, et al., *Risk Assessment,* p. viii.

21. For a discussion on what techniques are used by the industry, see Saul Levine, "The Role of Risk Assessment in the Nuclear Regulatory Process," *Nuclear Safety* (September/October 1978):556–564.

22. Rogovin, et al., *Three Mile Island,* p. 96.

23. Rasmussen, et al., *Reactor Safety Study.*

24. Nuclear Regulatory Commission and Environmental Protection Agency, *Planning Basis for the Development of State and Local Government Radiological Emergency Response Plans in Support of Light Water Nuclear Power Plants* (Washington, D.C.: Nuclear Regulatory Commission, 1978), p. 47.

25. Nuclear Regulatory Commission and Environmental Protection Agency, "Planning Basis for Emergency Response to Nuclear Power Reactor Accidents," NRC/EPA Task Force Report, NUREG 0396/EPA–5206–1–78–016; *Federal Register* 44 (23 October 1979):61123.

26. For a discussion of the handling of the TMI accident, see Daniel Martin, *Three Mile Island: Prologue or Epilogue?* (Cambridge, Mass.: Ballinger, 1980).

27. Rogovin et al., supra *Three Mile Island,* p. 133.

28. Federal Emergency Management Agency, *State Radiological Emergency Planning and Preparedness In Support of Commercial Nuclear Power Plants,* Report to the President (Washington, D.C., 1980), p. 1–3.

29. National Academy of Public Administration, *Major Alternatives for Government Policies, Organizational Structures, and Actions in Civilian Nuclear Reactor Emergency Management in the United States* (Washington, D.C.: Nuclear Regulatory Commission, 1980), pp. 65–66.

30. See Rogovin, et al., *Three Mile Island.*

31. U.S. General Accounting Office, *Areas around Nuclear Facili-*

ties Should Be Better Prepared for Radiological Emergencies (Washington, D.C., 1979), pp. 1, 28.

32. Ibid.

33. National Governors' Association, *Final Report of the Emergency Preparedness Project* (Washington, D.C.: Government Printing Office, 31 December 1978).

34. Federal Preparedness Agency, "Federal Response Plan for Peacetime Emergencies," (Interim Guidance) (Washington, D.C.: General Services Administration, 1976).

35. National Academy of Public Administration, *Major Alternatives,* p. 71.

36. Ibid. p. 82.

37. Rogovin, et al., *Three Mile Island,* pp. 134–136.

38. See Nuclear Regulatory Commission, "Regulatory Guide 1.101: Emergency Planning for Nuclear Power Plants," 10 (Washington, D.C., March 1977), p. 50.

39. *Northern State Power Company* v. *State of Minnesota,* 447 F.2d 1143 (8th Cir., 1971).

40. Nuclear Regulatory Commission, "Supplement No. 1, NUREG 75/111," (Washington, D.C.: Nuclear Regulatory Commission), 15 March 1977.

41. National Academy of Public Administration, *Major Alternatives,* p. 19.

42. See Kemeny, et al., *Presidential Commission on Accident at Three Mile Island.*

43. Nuclear Regulatory Commission, "Planning Basis for Emergency Response to Nuclear Power Reactor Accidents," Policy Statement, SECY–79–461 (Washington, D.C.: 5 October 1979).

44. Office of White House Press Secretary, "Fact Sheet: The President's Response to the Recommendations of the President's Commission on the Accident at Three Mile Island," 7 December 1979. Executive Order No. 12202, *Federal Register* 45 (20 March 1980):17939.

45. Federal Emergency Management Agency, "State Radiological Emergency Planning and Preparedness in Support of Commercial Nuclear Power Plants, Report to the President" (Washington, D.C., 1980).

46. F.C. Olds, "Emergency Planning for Nuclear Plants," *Power Engineering* 85 (August 1981):48–56.

47. Nuclear Regulatory Commission, "Beyond Defense-in-Depth: Cost and Funding of State and Local Government Radiological Emergency Response Plans and Preparedness in Support of Commercial Nuclear Power Stations," (Washington, D.C., Office of State Programs, 1980).

48. National Academy of Public Administration, *Major Alternatives,* pp. 25–31.

49. See "Nuclear Regulatory Commission, 10 CFR Parts 50 and 70, Emergency Planning, Final Regulations," *Federal Register* 45 (19 August 1979):55402; and the Nuclear Regulatory Commission and Federal Emergency Management Agency, *Criteria for Preparation and Evaluation of Radiological Emergency Response Plans and Preparedness in Support of Nuclear Power Plants* (Washington, D.C.: Federal Emergency Management Agency, 1980).

50. *Federal Register* 45 (19 August 1979):55404.

51. See Olds, "Emergency Planning for Nuclear Plants," pp. 50–51.

52. National Governors' Association, *Final Report.*

53. Nuclear Regulatory Commission, Office of State Programs, Information Report on State Legislation, "State Laws: Radiological Emergency Planning and Preparedness for Nuclear Power Stations," 31 December 1980; and "State Laws and Pending Legislation: Radiological Emergency Planning and Preparedness," 1 June 1981.

54. Federal Emergency Management Agency, "National Radiological Emergency Preparedness/Response Plan for Commercial Nuclear Power Plant Accidents (Master Plan)," *Federal Register* 45 (23 December 1980):84910.

55. Olds, "Emergency Planning," p. 56.

56. John C. DeVine, "A Progress Report: Cleaning up TMI," *IEEE Spectrum,* (March 1981):44–49.

57. Rogovin et al. *Three Mile Island,* p. 112.

58. Eliot Marshall, "Reactor Safety and the Research Budget," *Science* 214 (13 November 1981):766–768. For safety questions caused by aging light-water reactors, see Matthew J. Wald, "Nuclear Accident Raises Doubt on Safety Margins," *The New York Times,* 6 December 1981, p. 44.

59. Rogovin et al., *Three Mile Island,* p. 112.

60. See, for example, Thomas H. Pigford, "The Management of Nuclear Safety: A Review of TMI after Two Years," *Nuclear News* (March 1981):41–48.

61. See General Accounting Office, *The Nuclear Non-Proliferation Act of 1978 Should Be Selectively Modified* (Washington, D.C.: Government Printing Office, 21 May 1981).

62. See Alvin M. Weinberg, "Social Institutions and Nuclear Energy," *Science* (1972):276; and Weinberg, "The Future of Nuclear Energy," *Physics Today* (March 1981):48–56.

11 Information Policies and Computer Technology: The Case of Criminal-Justice Records

Steven W. Hays,
Donald A. Marchand, and
Mark E. Tompkins

Computer Technology and Public Policy

The growth of organizational record keeping and the use of information technology, especially computers, over the last twenty years has resulted in a number of efforts in the public and private sectors to develop national information systems. In the last decade in particular, several major proposals were initiated to take advantage of what Anthony Oettinger has called the "compunications" revolution, where data processing and communications technologies have converged.[1] The proposed development of national networks for the use of criminal-history records, EFT, and electronic-message systems directly reflects the linkage between technological possibility and organizational feasibility.

Since several of the national information systems currently under consideration contain the personal records of millions of Americans, a good deal of the public debate concerns the advisability of developing such networks and their possible implications for individual rights and liberties such as privacy and due process. How will individual rights and liberties be affected by national computerized information systems? What are the appropriate levels of regulation, oversight, and accountability for these networks? What is the appropriate role of the federal government vis-à-vis state and local governments and the private sector? In the 1960s and 1970s, such questions emerged and steps were made toward defining the nature of individual rights vis-à-vis organizational record keeping. In the 1980s, government and private organizations will be increasingly involved in policy debates over the types and levels of privacy safeguards and regulations that may be applied to national computerized information networks.

Our aim here is not to review and discuss the policy debates over balancing interests in personal privacy and computer technologies. Rather, we explore how information concerning the actual and potential adverse impact of national computerized information networks has been identified, assessed, and used in the policy debates. More specifically, our study of the use of technology assessments cum policy analysis in debates and decision making on national computerized record-keeping systems suggests several important findings:

> The types and amount of present information about the impact of organizational record keeping on individuals is inadequate as the basis for comprehensive information-policy formation.

> A number of significant constraints reduce decision makers' reliance on such information in policy deliberations.

> Unless greater information on the impact of computer technologies is generated and used earlier in the development of these networks, it will be difficult (if not impossible) to weigh adequately concerns with protecting individual rights such as privacy and due process against countervailing organizational goals and objectives.

These findings are based on, and illustrated by, a case study of the Department of Justice's efforts to develop a nationwide network of computerized criminal-history (CCH) records. The debate over CCH records is important for two related reasons. First, the debate over the development and use of a CCH data system has raged for nearly two decades. This protracted debate has highlighted the various issues and concerns that arise when state and federal governments endeavor to develop regulatory frameworks for ensuring privacy and security safeguards. Second, despite the rather extensive attention to the basic policy and management issues, the crucial questions concerning the impact of a nationwide arrest and conviction data system on individual rights and liberties remain unresolved. Thus, the CCH-policy debate represents a significant stage in the continuing evolution of national information systems and attendant public-policy debates.

The Development of Criminal-History-Record Use:
A Historical Perspective

Information about the impact of criminal-history-reccrd use has been developed over the last fifty years. The process of gathering and assessing information has been ad hoc rather than systematic and has closely followed the development of institutional linkages between local, state, and federal law-enforcement and criminal-justice agencies.

The evolution of modern criminal-history record-keeping practices has been directly tied to the movement toward police professionalism.[2] During most of the nineteenth century, informality and archaic management procedures characterized most police record keeping. If maintained at all, criminal-history records were composed of whatever information police officers chose to retain at the time of arrests. Because no regularized procedures existed for identifying offenders and transmitting information, most criminal records were used only within individual police jurisdictions.

As modern police departments began to develop, record-keeping practices rapidly improved. Much of this progress can be attributed to several important societal and technological developments. With the establishment of the first centralized police departments in the mid-nineteenth century, precinct-level and departmentwide reporting and record-keeping systems were devised to provide police executives with accurate evaluative and management data. The problems inherent in collecting accurate offender information were greatly alleviated in the early twentieth century with the development of the fingerprinting system. This system provided criminal-justice officials with an extremely reliable method of identifying suspects and criminals, and it served as a major impetus in the evolution of criminal-history data systems. A final catalyst for upgrading the quality and comprehensiveness of criminal records lies in changes in U.S. society. As citizens became more mobile with the expanded use of automobiles and mass transportation, police departments encountered severe difficulties in tracking offenders between and among jurisdictions. Consequently, pressures came from law enforcement for the development of more-centralized and -institutionalized means of information collection and exchange.

After several aborted attempts to establish a national criminal-records repository, the Division of Identification was created within the FBI in 1925.[3] The division incorporated existing files that had earlier been collected by other state and federal officials. The FBI also thereby acquired visibility and offered criminal-justice information to state and local police. The FBI subsequently strove to increase the number of files maintained and available to law-enforcement officials. By 1943, for example, the FBI reportedly housed over 76 million civil- and criminal-identification records, as compared to only 2 million when the division was first established, less than twenty years earlier.

The movement at the federal level to develop a national clearinghouse for identification and arrest records was paralleled during this period by efforts to establish state-level repositories for identification and arrest records. Indeed, by 1931, forty states had established bureaus of identification.

With the appearance of second- and third-generation computer technology in the early and mid-1960s, the FBI, in cooperation with state and local law-enforcement agencies, began experimenting with automated files.

In 1967, the National Crime Information Center (NCIC) became operational with approximately 23,000 records of wanted persons and stolen property in its computer files. By the late 1960s, a prototype computerized exchange of criminal-history information was successfully tested. In December 1971, after prolonged debate within the Department of Justice, the FBI assumed control of the CCH program. Then, in July 1973, the FBI requested permission to expand the communications capability of the NCIC to include message switching for NCIC files, including CCH. For the next five years, a heated policy debate focused on this issue. However, in 1978, the Department of Justice proposed a consensus concept for a nationwide criminal-justice-information-interchange facility, and the FBI proceeded with a long-range plan to automate its identification and arrest records contained in the Identification Division.[4] The development of CCH records was also furthered by the Law Enforcement Assistance Administration (LEAA). The LEAA in 1972 began a comprehensive data-systems program to help fund state and local criminal-justice information systems. Within five years, the LEAA had also identified over 683 separately defined criminal-history-record systems representing 549 jurisdictions. The Department of Justice and state and local law-enforcement organizations continue to promote the development of the nationwide exchange of criminal-history information.

Studies suggest that at least 37 million—but the figure may run as high as 50 million—people have arrest records.[5] The number of living persons with arrest records in FBI files is estimated to be more than 16 million. Also, another 216 million criminal records are in existence in criminal-justice agencies throughout the United States. Of this number, 21 million are held by the FBI and 196 million are held by more than 56,000 state and local criminal-justice agencies, of which about 29 million are in state repositories and the rest in local criminal-justice agencies.[6]

Given the availability of this vast number of fingerprint-based records, other governmental uses of identification files began to appear. As early as the 1930s, the government began to use identification and arrest records for public-employee background checks. Between 1930 and 1960 this trend continued at a rapid rate. Congress, moreover, gave the FBI additional responsibilities for conducting background checks on individuals in order to determine their suitability and general fitness for both public and private employment. Additionally, the use of identification records to ensure political loyalty and good moral character for licensing purposes became widespread at both the state and federal levels of government.

In the 1940s and 1950s, the ballooning use of criminal-history records for non-criminal-justice purposes went unchallenged. A few organizations such as the American Civil Liberties Union (ACLU) and labor unions openly expressed concern about the scope of domestic security and loyalty programs and the use of identification files in strike-busting efforts. These protests were nonetheless relatively infrequent and ineffective.

By contrast, during the 1960s, concerns about the individual and social impact of criminal-history records developed within the context of a wider policy debate over privacy and computers. In the early 1960s, popular and scholarly attention began to focus on the possible adverse impacts of computers. Still, not until the National Data Center controversy (that is, a series of very visible and contentious Congressional hearings in the mid-1960s concerning whether or not the federal government should merge its statistical record-keeping files) did the privacy-and-computers issue receive national attention as the subject of congressional and political debates. Introduction of third-generation computer technology in turn heightened public awareness about the use of computers. Critics of widespread and unaccountable use of computer technology also had an opportunity to challenge public and private organizations on two fronts. First, they pointed to concerns about the centralization of files and the ease of access to personal-record information provided by computers. Second, they questioned the standards and policies adopted by organizations in making decisions about individuals. Throughout the 1960s these concerns gained visibility as the public and government officials became more cognizant of the limitations of existing information policies and practices. From 1965 to 1970, numerous legislative proposals for regulating record-keeping systems were advanced.

The concern in the 1960s over privacy and computers nevertheless did not directly affect the development of the NCIC. Throughout the decade, the NCIC was perceived as a "thing" file with little identifiable information on individuals except those who had outstanding warrants. Criticism of criminal-justice record keeping was primarily directed at the rap sheets in the FBI's Identification Division. The ACLU, for example, continued to attack the lack of accuracy and completeness in the FBI's handling and dissemination of arrest records. In June 1969, at its biennial conference, the ACLU put itself on record as a "public spokesman in the defense of personal privacy and civil liberty" in the wake of the National Data Center controversy:

> Whenever a government amasses files about its citizens an inherent threat to liberties exists. . . . The National Data Bank proposals exemplify such use; the seeming insensitivity of its proponents for safeguards underscores the need for legislative protections. . . . The ACLU believes that the process of converting manual records to computer processing poses a great risk to privacy and due process.[7]

In the 1970s, further development of proposals for automating criminal-history records heightened concerns about their use and potential abuse.

Almost from the start of the privacy-and-computers debate, the use of criminal records for other than criminal-justice purposes had been singled out as an important example of the problems posed by computer technology. Thus, proposals for an integrated network of criminal-information systems and an interstate criminal-history exchange system did include some safeguards against unauthorized use of criminal-history records. Still, at this stage of the public-policy debate, a consensus emerged only on the fact that the uses of criminal-history records had to be regulated if advanced criminal-justice information systems were to be established. There was no agreement on the nature and extent of governmental regulation and privacy protection safeguards.

With the growth of LEAA-funded state and local criminal-information systems and the implementation of the FBI's CCH program, principal concern shifted from the potential adverse effects of computers to the need for more-effective and systematic information policies governing the collection, maintenance, and dissemination of criminal-history records. Concomitantly, formulation of information policy moved from an informal and abstract stage to the center of the policy debate over the regulation of criminal-justice information systems. Two principal issues of controversy emerged: (1) How would the collection and exchange of criminal-history information be managed and controlled; and (2) what role would the LEAA assume in formulating and regulating state and local criminal-history-information policy?

By 1974, these two issues were nonetheless again merged in policy disputes over proposals to develop a comprehensive national policy for the use of criminal-history records. This struggle resulted in a number of attempts to pass legislation, but congressional efforts were again frustrated and met with resistance from the criminal-justice community, the press, and special-interest groups. In particular, during 1975, the controversy over message switching erupted into a full-scale bureaucratic debate between the LEAA and the FBI but also involved the states, public interest groups, and congressmen intent on assuring that individuals' privacy rights were properly protected. As a result of protracted debates in the mid-1970s, legislative initiatives to achieve consensus on criminal-justice-information policy lost momentum.

In the 1980s, criminal-justice-information policy has continued to be characterized by intense organizational conflicts at the federal level and congressional oversight of the future of the CCH program and message switching by the FBI. While state and local governments have made some progress in developing legislative and administrative controls for the use of criminal-history records, the federal government continues on a course that can only be characterized as one of misunderstanding, polarization, and indecision.

**Limitations of Information about the Impact of
Criminal-Record Use**

Despite the long history of criminal-record use in the United States and the
intense policy debates in the 1970s concerning privacy and security safeguards applicable to criminal-history records, no clear picture of either the
precise number of individuals affected or the consequences and costs of
criminal-history-record use has emerged. As late as 1978, the OTA, after
examining the problems related to the CCH program, concluded that no
reliable information existed concerning the nature and extent of the impact
on individuals of criminal-record use on which to base the formulation of
information policy.[8] While the OTA did not account for such deficiencies in
information, it did raise an important question regarding the conduct of the
policy debates over national information systems in the absence of reliable
information concerning the individuals affected.

The major types of information concerning the impact on individuals
of criminal-record use is, as the following list shows, quite broad:

    Anecdotal evidence:
    Court cases,
    Individual complaints to:
      Legislative committees,
      Public interest groups,
      Public organizations.

    Empirical evidence:
    Estimation studies of populations affected,
    Examinations of record use in specific decision-making contexts:
      Within criminal-justice process:
        By police,
        By prosecution,
        By probation,
        By corrections,
        By ex-offender self-help groups,
        By offender rehabilitation organizations;
      Outside criminal-justice process:
        By public employers,
        By licensing agencies,
        By social-service agencies,
        By military,
        By private employers,
        By credit agencies,
        By banks,

By private-investigation firms,
By insurance companies,
By bonding agencies,
By the press;
Examinations of record transactions by law-enforcement and criminal-justice record keepers.

Arrest and conviction information continues to be used by both public- and private-sector organizations. Yet, information about the nature of the impact on individuals of criminal records remains quite uneven.[9] The history of policy debates over criminal-record use, for instance, reveals little hard evidence on individual complaints, and relatively few court cases have involved individuals adversely affected by record-use discrimination.[10] Moreover, there have been no separate investigations of how often criminal-justice agencies or other public agencies receive and respond to complaints. Even though considerable testimonial evidence exists from public interest groups and legislative committees, showing that large numbers of individuals complain about record systems, the number and nature of the complaints are rarely documented.[11]

Most of the evidence of computer record systems' problems has derived from a small number of empirical investigations. Several studies have aimed at defining the number of individuals with arrest or conviction records in the United States.[12] Additionally, throughout the 1960s and 1970s, studies by sociologists and political scientists surveyed the operating policies and practices of selected public and private organizations that use criminal-history information in their decision-making processes.[13] Study commissions and committees at the state and local level have also taken testimony from public and private organizations about the extent and nature of individual record use.[14] Normally, however, proponents and opponents of privacy and security policies merely refer to so-called record problems that individuals experience without documenting those or exploring other potential problems.[15] Finally, most of the major studies of criminal-record information are now seriously outdated. Many of the studies were completed between 1960 and 1975 and, thus, neither accurately describe record policies and practices in the 1980s nor adequately sample the populations affected.[16]

Two further factors have inhibited assessments of how information contained in criminal-record repositories is used. First, obstacles are presented by the sheer volume of records and their numerous uses by private and public organizations. For instance, in one state, over 150 secondary users, 50 federal users, and 3,000 criminal-justice users had access to state and federal criminal-history files through approximately 2,000 terminals.[17] Second, actual use by organizations of record systems is an amorphous, problematic concept. Actual use can only be studied by scrutinizing the internal workings of agencies and firms. Such research is usually difficult to perform and demands attention to qualitative as well as quantitative factors. Measures of use that are presently available—like transactions derived from record-repository reports—inadequately indicate actual use of record systems since this information is aggregated at the level of the system or unit and therefore does not suggest how the information was in fact used in an organization's decision-making process.

In sum, the level and nature of present information about the use of computerized record systems is a poor basis on which to assess the individual and social impact of a national information system. Furthermore, as we argue in the next section, a variety of constraints continue to exist in the policy process that makes efforts to develop more-systematic and -reliable information difficult at best.

Constraints on the Identification and Evaluation of the Impact of Criminal-Record Use

The identification and evaluation of the individual effects of record use by public and private organizations is not simply an information problem. Instead, a number of significant questions arise about the political and organizational factors that tend to reduce incentives for policy actors to expend resources (time and money) to gather pertinent information and to use that information in public-policy formulation. A number of the most important disincentives for information collection and analysis, as well as constraints on its usage in public policymaking, are examined here.

Varying Perceptions of the Importance of Impact Information

Four major views of the value of impact information emerged during the policy debates over CCH records in the 1970s. These divergent perceptions divided the professional and academic communities and thereby inhibited the consensus building that must precede formulation of public policy.

Those in the law-enforcement community interested in developing automated information systems tended to view the issue of the possible impact of criminal-record uses as not really very significant. In their view, if there were any adverse impact of such systems, criminal-justice agencies would have already responded. Moreover, neither had individuals complained in any great numbers about the impact of computerized records nor had public interest groups like the ACLU actually identified widespread abuses and problems.[18]

Some academics and scholars in the privacy-and-computers field agreed but also deflected attention away from the issue of criminal-justice records. Alan Westin and Michael Baker, for example, suggested that automation did not really increase any negative impact on individuals of record keeping.[19] The same policies and practices that were applied to automated records had previously governed manual records. Automation in and of itself, they argued, neither had nor would have serious negative consequences for individuals. Accordingly, computer technology was simply viewed as an appropriate tool for achieving societal goals like crime control.

By contrast, some congressmen and special-interest groups viewed the impact of criminal-record systems as serious and pervasive and maintained that no one could identify those impacts or their social costs precisely. In their view public policy responded to, and was based on, an extrapolation of the problems posed by court cases and complaints of selected individuals.

Finally, some critics argued that we really know neither what the nature, extent, and consequences of record-keeping practices are nor whether they can be precisely identified and quantified. However, before developing large-scale information networks, they urged a systematic investigation of those possible impacts. In the late 1970s, such investigation was (and not surprisingly) advocated by the OTA.

Issue Characteristics

Despite the attention given to privacy and computers as a public-policy issue, the plight of individuals adversely affected by arrest-and-conviction-record use remain a low-visibility, and diffuse concern.[20] The low visibility and intensity of criminal records for the public and the attendant problems of identifying and evaluating the impact of record systems on individuals nonetheless has several crucial policy implications.

The intensity of a public-policy issue depends on public perception of its importance. Public perceptions, of course, shift over time. Given the low intensity of public interest in the issue (except among those with a direct stake in the issue, as with individuals with criminal records), it becomes difficult even to represent this issue in the policymaking process. Low intensity means low visibility in the political process; hence, dissemination of information about the impact of developing technologies becomes extremely important.

Another related problem involves the difficulty of representing the interests of individuals in the policy process: "The definition of [an] interest is inextricably bound up with the conception of what constitutes protection of that interest."[21] Both the FBI and the ACLU, for example, may claim to represent the interests of individuals whose names are included in criminal-justice information systems. Their respective claims, however, depend on their perceptions of the nature of the problems attending the use and dissemination of criminal-history records. Even though they have nearly opposite approaches to the problem of criminal-justice records, both groups can claim to be the true representatives of individuals' interests. Assessment of the impact of computerized record systems thus would not only permit an informed evaluation of the consequences for individuals of such systems but also would permit policymakers to make more-intelligent judgments and to define more perceptively the policy issues and trade-offs.

Mixed Organizational Incentives

Public agencies ordinarily have little or no incentive to reduce the social costs attendant their maintenance or operating costs.[22] In other words, agencies that are neither directly nor indirectly responsible for assessing the. impact of their operating practices have little incentive to represent outside interests. However, when the social impact of an agency's internal operating practices is serious enough to threaten its continuance, or simply when the financial costs are less than the political benefits of minimizing adverse social impacts, agencies can be expected to have some representational role and to reduce the adverse impacts arising from their maintenance processes. Within the context of CCH records, criminal-justice agencies have indeed had selective incentives to develop information policies and to reduce the adverse impacts of criminal records.

Concern for the impact of criminal-history records on individuals, as discussed earlier, developed as part of the wider policy debate over balancing interests in privacy and computerized information systems. The National Data Center controversy remains an important lesson for proponents of criminal-justice information systems: If the problems of privacy and other social costs presented by proposed information systems are not recognized and evaluated at the outset, public criticism and legislative inquiries could subsequently force curtailment of the proposed projects. Criminal-justice agencies, including Project SEARCH—the national consortium of state criminal-justice officials originally responsible for developing the CCH prototype for the LEAA (now known as SEARCH Group, Inc.)—realized the need to review the uses of criminal-history records if they were to pursue their goals of developing advanced criminal-justice information systems at the local, state, and federal levels.

Much of the representational role of criminal-justice agencies in information-policy formulation was obviously linked to their goal of developing computerized information systems. Indeed, as public interest groups and the press highlighted the problems of privacy and computers, criminal-justice agencies increasingly acted to ensure some safeguards for the dissemination of information. Moreover, and not surprisingly, criminal-justice agencies are also beginning to take the initiative in developing information policies that suit their needs and plans rather than relying on a reactive posture. Particularly since 1975, the promotion of policies responsive to the concern for privacy, confidentiality, and security of criminal records has represented good politics for federal agencies. In fact, a large measure of the LEAA's ability to maintain its own position on the CCH program and message switching, when confronted with intensive FBI lobbying in the Justice Department and Congress for its own record system, derived from the LEAA's active support for privacy-and-security concerns before sympathetic members of Congress and outside interest groups. The FBI, however, has been viewed as being less receptive to privacy concerns, and therefore its proposals have encountered resistance by members of Congress who simply do not trust the FBI to protect citizens' privacy rights.[23]

Although the assumption of an active posture vis-à-vis privacy-and-security concerns helped the political standing of SEARCH Group Inc. and the LEAA, it has not led them to develop new information concerning the nature and extent of the impact of criminal-history-record systems on individuals. Many of the guidelines, recommendations, standards, and suggested laws advocated by these groups developed without systematic identification and evaluation of the social impact of these systems. Instead, they relied on perceptions of criminal-justice practitioners about the impact of record systems and the ways those impacts could be addressed in the process of developing fully automated information systems.

Judicial Constraints

The judiciary has played an important role in the information-policy arena. The import of the courts lies not so much in the remedies granted to individuals adversely affected by criminal-history records but because they provide an alternative forum in which public interest groups may pursue their policy preferences through test cases. In other words, the political significance of cases such as *Menard* v. *Mitchell* and *Menard* v. *Saxbe* derives less from the particular relief granted than from the discovery and dramatization of deficiencies in existing information practices and thus the need for legislative action.[24] The judicial forum enables representation of nonintense, diffuse interests as well as permits groups and individuals to extend the meaning and reach of rights such as privacy, thereby forcing public and private organizations to internalize the social costs of computerized record systems. In recent years, courts also provided an important (albeit limited) source of factual information about the connection between social costs of criminal-justice records and the practices and policies of criminal-justice agencies. Not only have court cases revealed evidence of specific abuses, but they have also urged more-extended fact finding concerning the scope and nature of record-keeping activities and their negative social and economic effects.[25]

Disincentives for Collective Action

The segment of the public directly and potentially adversely affected by uses of criminal-history records has more disincentives than incentives for taking collective action against the misuse and abuse of criminal-justice records.[26] According to Mancur Olsen, separable and selective incentives are needed to stimulate individuals to act as a group in order to achieve some collective good.[27] The presence or absence of incentives affects the likelihood of a collective good being provided to a group. When we consider that portion of the public most directly affected by the individual costs of criminal-history records, however, it becomes clear not only that those individuals do not enjoy separable and selective incentives for collective action but also that selective disincentives actually operate to encourage collective inaction.

The main impact of criminal-history records on individuals lies in the closing of employment opportunities. Individuals with arrest or conviction records have difficulty in obtaining employment in private industry, public service, and licensed occupations. They also encounter difficulties in obtaining loans, insurance, credit, public housing, and commercial bonding for selected occupations. In addition, individuals may be rejected by or separated from the armed forces because of their arrest or conviction records as well as become targets of private investigation that may adversely affect their legitimate economic, political, or social interests.

The cumulative effect of these uses of criminal-history records on individuals inclines them neither to talk about nor bear witness to the arrest or conviction record except under special conditions. An arrest or conviction record constitutes a liability or a stigma that individuals must respond to by managing such information about themselves in a manner that reduces the personal costs that are likely to be incurred. Individuals stigmatized by such records have not organized precisely because each individual seeks to conceal and thus to escape the consequences of his criminal-history record. Only when individuals have records that are most difficult to conceal—for example, as with criminal convictions—have they joined together to pursue their collective interest in limiting the use of criminal-justice records. In recent years, ex-convicts formed a number of organizations to concentrate public attention on their plight and to help each other obtain suitable housing and employment.[28] However, notably, such organizations have not had memberships commensurate with the actual number of eligible individuals. Moreover, many of their associations comprise what Olsen terms "forgotten groups—unorganized groups having only limited resources to support political action on their behalf and either no lobby or a very weak lobby to promote and protect their interests.[29] Given that the portion of the general public affected by the social costs of criminal-history records is composed of large segments of the young, the unemployed, the poor, and the black and other racial minorities, it is apparent why such individuals are unorganized and not stimulated to act collectively. Traditionally, such groups have not received the immediate benefits and protection derived from collective action, and thus they often remain the losers in the political process.

Obscuring Symbolic Factors

In recent years, the right of privacy has become politically symbolic. The recognition of the import of personal privacy, in view of increased organizational record keeping and computerizaton, has rendered privacy not only a concrete right to be reconciled with competing social goals but also a political resource and symbol. Affirmation of this right in political debates lends legitimacy to what are basically political preferences.[30] In other words, the introduction of an abstract right of privacy in U.S. politics has tended to symbolize political reforms. As Stuart Scheingold states, "It is possible to capitalize on the perceptions of entitlement associated with rights to initiate and to nurture political mobilization—a dual process of activating a quiescent citizenry and organizing groups into effective political units."[31]

Privacy as a political symbol served a useful purpose in political debates over the development and regulation of computerized information systems. However, since many separable issues tend to be included in concerns for privacy protection, discussion of privacy protection has often obscured and overshadowed the more-specific questions and problems related to the effects on individuals of arrest-and-conviction-record use. More exactly, claims for privacy and security regulations tend to obscure the fact that other social values—including due process, equal protection, and liberty—are at stake in assessments of criminal-justice-record practices. In sum, the rhetoric of privacy protection served as a convenient shorthand for a whole range of individual and social concerns about organizational record keeping but also tended to downplay the need for systematic assessment of record uses, impacts, and policy trade-offs.[32]

Practices of Professional Record Keepers and
Information Technologists

The identification and evaluation of the impact of criminal-record use is intimately dependent on information concerning the use and management of these records by professional record keepers and information specialists. A recent study of the uses of criminal-record information in interstate systems found that it is difficult, if not impossible, to trace the uses of records in many information systems.[33]

Most criminal-information systems were designed for one purpose—namely, providing a service to users. Accordingly, few data systems are capable of generating statistics on other than aggregate message traffic to various automated files. The NCIC, for example, only records aggregate traffic to and from the sending and receiving terminal. Because state criminal-record repositories normally do not maintain terminal-by-terminal use statistics, information dissemination beyond the state repositories remains somewhat of a mystery. Moreover, even when such records are kept, they are not very useful for one major reason: As many as 100 different users may have access to any given local terminal (for example, state agencies, probation departments, courts, licensing boards, and so on). Thus, little is actually known about use of individual criminal histories, and it is very costly and time consuming to reach individual users, who may number in the thousands, to create an accurate picture of criminal-record-information use.

Toward Improving Assessments of the Impact
of Computerized Record Systems

Identification and evaluation of the effects of organizational record-keeping practices remains crucial for the process of developing national information systems. Still, pervasive obstacles also exist to assessing the impact of computer technologies on individuals. Moreover, endemic to this policy area is a lag between the push to automated national information systems and systematic assessments of the individual and social impact of computerized records. If systematic assessments are to keep pace with managerial and technological advances, then some basic changes are needed in the prevailing approaches to planning, supporting, and evaluating large-scale information networks.

Agencies directly involved in the planning and promoting of national information networks should be legally required to conduct systematic assessments of the individual and social impacts of their record-keeping and use practices. Assessment of individual and social impacts of national information systems should become part of the technical, organizational, and management analysis undertaken prior to the development of national information systems. Individual- and social-impact assessments should be an integral part of the efforts of agency planning and decision-making processes. In other words, agencies should bear a burden of proof in justifying that their proposed information systems will not unduly intrude on the rights and liberties of citizens. Ostensibly, individual- and social-impact assessments would enhance the public accountability of computerized record systems.[34]

National information systems, moreover, should not be initiated without statutory guidelines.[35] While Congress has played a vital role in the policy debates over criminal-history information networks, most of its legislative proposals responded to executive-branch actions to develop record systems. Hence, congressional committees and staffs have found it difficult to respond adequately to well-organized agency efforts to develop information networks and, at the same time, to assess the potential impact of those networks on individuals. Additionally, agencies have not been required to develop consistent and comprehensive system-development plans, and thus their uncoordinated policy initiatives work at cross purposes and force Congress to respond to a chaotic and conflicting array of executive-branch proposals for the development of a national information network. In this regard, Congress should make its main technology oversight agency, the OTA, responsible for regular and periodic reviews and evaluations of the development of national information systems instead of requesting the OTA to undertake one-time reviews of computerized record systems.[36]

Finally, adequate funds should be provided to the executive branch for undertaking basic and applied research into the assessment of the individual and social impact of national computerized information systems. As suggested earlier, a much larger proportion of agency funding is directed at the development and promotion of information technology and systems development than at assessment of individual and social impacts. Neither the NSF nor agencies like the LEAA have directed sufficient resources to the systematic assessment of the impacts of national information systems. While the NSF supports basic and applied research in telecommunications, computer science, and other areas of technology assessment, its funding programs have not specifically aimed at the improvement of approaches and methods for individual- and social-impact assessment.[37]

Conclusion

Improving knowledge about the impact of large, complex national information systems seems essential to the fundamental political choices underlying technology-related public policies. Construction of vast networks of information systems, maintaining personal records on millions of individuals, has proceeded without assessing the long-term and possible irreversible effects of those systems on individuals in our information-oriented society. Assessments of the impact of technological innovations seems crucial if policymakers are to evaluate the external and intangible social costs of developing technologies and to decide intelligently whether and to what extent regulations are necessary for the acquisition, maintenance, and dissemination of personal information contained in the ever-expanding banks of computerized record systems.

Notes

1. See Anthony Oettinger and John C. Legates, *Domestic and International Information Resources Policy* (Cambridge: Program on Information Resources Policy, Harvard University, 1977), pp. 13–31.

2. For an extensive review of the history of law-enforcement record-keeping, see Donald A. Marchand and Eva G. Bogan, *A History and Background Assessment of the National Crime Information Center and Computerized Criminal History Program,* Contractor Report. (Washington, D.C.: Office of Technology Assessment, June 1979), pp. 6–34.

3. See Donald C. Dilworth, ed., *Identification Wanted* (Gaithersburg, Md.: International Association of Chiefs of Police, 1977).

4. For a brief description of the efforts of the Identification Division to develop an automated fingerprint-scanning and arrest-record system called AIDS, see Steven W. Hays, Eva G. Bogan, and Donald A.Marchand, *An Assessment of the Uses of Information in the National Crime Information Center and Computerized Criminal History Programs,* Contractor Report (Washington, D.C.: Office of Technology Assessment, October 1979), pp. 23–38.

5. For a comprehensive assessment of such studies, see Lynne Eckholt Cooper, Mark E. Tompkins, and Donald A. Marchand, *An Assessment of the Social Impacts of the National Crime Information Center and Computerized Criminal History Program,* Contractor Report (Washington, D.C.: Office of Technology Assessment, October 1979), pp. 51–84.

6. See Kenneth C. Laudon, "Privacy and Federal Data Banks," *Society* 52 (January/February 1980):52.

7. "Minutes of Due Process Committee, 10 March 1960," in *Policy Guide of ACLU, Policy* (New York, 1970), p. 202.

8. U.S. Congress, Office of Technology Assessment, *A Preliminary Assessment of the National Crime Information Center and Computerized*

Criminal History Program (Washington, D.C.: Government Printing Office, 1978), p. 16.

9. See Cooper, Tompkins, and Marchand, *Assessment of Social Impacts.*

10. Donald A. Marchand, *The Politics of Privacy, Computers and Criminal Justice Records* (Arlington, Va.: Information Resources Press, 1980), pp. 85–111.

11. See, for example, U.S. Congress, House, Committee on the Judiciary, Subcommittee No. 4, *Security and Privacy of Criminal Arrest Records,* Hearings, 92nd Cong., 2d sess. 16, 22, 23 March and 13, 26 April 1972; U.S. Congress, House, Committee on the Judiciary, Subcommittee of Civil and Constitutional Rights, *Criminal Justice Information Control and Protection of Privacy Act,* 94th Cong., 1st. sess., 14, 17 July and 5 September 1975; U.S. Congress, House, Committee on the Judiciary, Subcommittee on Civil Rights and Constitutional Rights, *Dissemination of Criminal Justice Information,* Hearings, 93rd Cong., 1st and 2d sess., 26 July, 2 August, 26 September, and 11 October 1973 and 26, 28 February, 5, 28 March, and 3 April 1974.

12. See R. Christensen, "Projected Percentage of U.S. Population with Criminal Arrest and Conviction Records," in *Task Force Report: Science and Technology,* President's Commission on Law Enforcement and Administration of Justice, ed., pp. 216–218 (Washington, D.C. Government Printing Office, 1969); and Neal A. Miller, *A Study of the Number of Persons with Records of Arrest or Conviction in the Labor Force* (Washington, D.C.: Department of Labor, January 1979).

13. See, for example, James W. Hunt, J.E. Bowers, and N. Miller, *Laws, Licenses and the Offender's Right to Work* (Washington, D.C.: National Clearinghouse on Offender Employment Restrictions, 1974); Herbert S. Miller, *The Closed Door: The Effect of a Criminal Record on Employment with State and Local Public Agencies* (Washington, D.C.: Manpower Administration, Department of Labor, 1972); and Herbert S. Miller and Marietta Miller, *Guilty But Not Convicted: Effect of an Arrest Record on Employment* (Washington, D.C.: Georgetown University Law Center, 1972).

14. See, for example, California Legislature, Assembly Select Committee on the Administration of Justice, *Security and Privacy and Criminal History Information Systems* (Sacramento: March 1971); and City of New York, Commission on Human Rights, *The Employment Problems of Ex-Offenders* (New York, 22–25 May 1972).

15. See, for example, National Advisory Commission on Criminal Justice Standards and Goals, *Report on the Criminal Justice System* (Washington, D.C.: Government Printing Office, 1973); Project SEARCH, *A Model State Act for Criminal Offender Record Information,* Technical

Memorandum No. 3 (Sacramento: California Crime Technological Research Foundation, 1971); and Project SEARCH *Model Administrative Regulations for Criminal Offender Records Information,* Technical Memorandum No. 4 (Sacramento, California Crime Technological Research Foundation, 1972).

16. See Miller, *Study of Persons with Records;* and Miller, *Employer Barriers to the Employment of Persons with Records of Arrest or Conviction* (Washington, D.C.: Department of Labor. 1979).

17. See Hays, Bogan, and Marchand, *Assessment of Uses of Information,* pp. 10–11.

18. Although local and state ACLU chapters have received many individual complaints about arrest- or conviction-record problems, the national office does not systematically assemble or analyze such information.

19. See Alan F. Westin and Michael A. Baker, *Databanks in a Free Society* (New York: Quadrangle Books, 1972).

20. See Marchand, *Politics of Privacy,* pp. 249–251.

21. Mark V. Nadel, *The Politics of Consumer Protection* (New York: Bobbs-Merrill, 1971), p. 237.

22. See Marchand, *Politics of Privacy,* pp. 260–262.

23. Ibid., p.262.

24. Ibid., pp. 182–184.

25. See *Tatum* v. *Rogers,* 75 Civ. 2782 (S.D.N.Y. 1977).

26. See Marchand, *Politics of Privacy,* pp. 108–111.

27. Mancur Olsen, *The Logic of Collective Action* (Cambridge: Harvard University Press, 1965), p. 33.

28. See Mary Lee Bundy and K.R. Harmon, *The National Prison Directory, Base Volume,* Supp. Nos. 1 and 2 (College Park, Md.: Urban Information Interpreters, 1975–1977).

29. Olsen, *Logic of Collective Action,* p. 165.

30. Stuart A. Scheingold, *The Politics of Rights* (New Haven: Yale University Press, 1974), p. 85.

31. Ibid., p. 131.

32. For a review of the complexity of social goals and policies involved in the assessment of criminal-history use, see Marchand, *Politics of Privacy,* pp. 85–104.

33. See Hays, Bogan, and Marchand, *Assessment of Uses of Information,* pp. 11–12.

34. For an assessment of the use of environmental-impact statements, see Council on Environmental Quality, *Sixth Annual Report* (Washington, D.C.: Government Printing Office, 1975), pp. 626–634.

35. See Laudon, "Privacy and Federal Data Banks," p. 55.

36. Although the GAO has provided reports on the CCH controversy, these have not focused on individual- and social-impact analysis. See U.S.

Comptroller General, *Development of a Nationwide Criminal Data Exchange System: Need to Determine Cost and Improve Reporting* (Washington, D.C.: General Accounting Office, 1973); "Development of the Computerized Criminal History Information System," Report to Senator Sam J. Erwin, chairman, Subcommittee on Constitutional Rights (Washington, D.C. General Accounting Office, 1 March 1974); *How Criminal Justice Agencies Use Criminal History Information* (Washington, D.C.: General Accounting Office, 1974).

    37. National Science Foundation, *Guide to Programs* (Washington, D.C.: Government Printing Office, 1979).

# 12

Technology, Society,
and Public Policy: EFT
Systems

Kent W. Colton and
Kenneth L. Kraemer

The Evolution of EFT: Major Forces and Values

In its final report of February 1977, the National Commission on Electronic Funds Transfer (NCEFT) defined EFT as "a payment system in which the processing of communications necessary to effect economic exchange, and the processing of communications necessary for the production and distribution of services incidental or related to economic exchange, are dependent wholly or in large part upon the use of electronics."[1]

This definition of EFT is inadequate in light of the emerging recognition that EFT is a complex, technologically based system with the potential for vastly changing relationships among private enterprise, public institutions, and individuals throughout the country. The complexity of EFT is illustrated by the fact that it is neither a single technological application nor even composed of a unified set of technological applications or services. At least eight different overlapping technologies and services characterize the applications being developed in this country under the rubric of EFT. EFT technologies can be divided into two broad areas: (1) centralized services and (2) partially decentralized services. Centralized services, usually initiated by financial institutions or governmental agencies on behalf of customers, are processed at main locations and then returned to the financial institution, primarily banks. Decentralized services are typically initiated by customers or merchants and involve the use of free-standing EFT terminals to initiate transactions from a variety of locations.

The principal technologies associated with EFT services are identified in table 12–1. They include automatic clearinghouses (ACH), direct deposits, bill paying, wire transfers, automatic-teller machines (ATM), point-of-sale (POS) terminals, check/credit-card authorization, and telephone transfers. Individually, and through a combination of these subsystems and technologies, EFT operating systems are being established in various areas throughout the country.[2]

Authorship is alphabetical to denote equal contribution. The chapter is based on current research by the authors and was supported by an OTA contract with the Irvine Research Corporation for assessing equity and privacy issues in EFT.

241

Table 12-1

Major EFT Facilities and Services Currently under Development

Subsystem or Techniques	Operation
Centralized Services (those initiated by the financial institution)	
Automated clearinghouses (ACHs): Financial-transaction centers that electronically process account transfers of member financial institutions	An electronic network(s) substitutes for the paper-oriented check-clearing system. Also provides the clearing facility for pre-authorization procedures and POS operations, but use is small.
Direct deposit: Prearranged routing of the individual income payments directly to personal accounts at financial institutions	Once authorized, such deposits and payments are made automatically and electronically according to agreed-upon procedures.
Bill paying: Preauthorized debiting of customer accounts at fixed periods to pay recurring bills	Customers authorize financial institutions to pay regularly monthly bills (usually through electronic transfers) or to transfer money from a savings to a checking account and vice versa.
Wire transfers: Instantaneous transfer of funds between financial institutions, Federal Reserve branches, and government offices	The two existing systems include Fedwire, managed by the Federal Reserve, and Bankwire.
Decentralized Services (initiated by the customer or merchant)	
Automatic-teller machines (ATM): Automatic terminals connected electronically to a financial institution that will perform most banking functions of human tellers	ATMS provide 24-hour banking service through electronic terminals; over 8,000 ATMs are in place today.
Point-of-sale (POS) terminals: Electronic terminals located in retail establishments to facilitate immediate exchange of funds and merchandise	Verify or guarantee a check electronically or make a direct, electronic debit from a purchaser's account to the account of a business establishment at the point of sale (for example, the so-called debit card).
Check/credit-card authorization: Immediate verification of sufficient funds or credit to purchase goods or services	A terminal is used to check the customer's credit and to determine whether the checking or credit accounts have adequate funds to handle the transaction in question.
Telephone transfers: Electronic transfers of customer funds through telephone instructions to computers or personnel at a financial institution	Customers authorize telephone transfers to pay specific bills on a decentralized basis.

The success, failure, and indeed, the very nature of EFT systems is nonetheless dependent on several forces surrounding the development of EFT. The major forces involved in the evolution and development of EFT systems are private and public institutional actors, EFT technology and operating systems, and monitoring and evaluating EFT systems.[3] EFT operating systems are only one part of the larger system of financial institutions and, hence, are inherently involved in the major public-policy and political questions that traditionally arise with regulation of the private sector.

The institutional actors involved in the provision and use of EFT technology play an important role in determining the overall strength of EFT operating systems. What consumers and users demand or will accept is a major determinant in what EFT providers—the financial institutions and retailers who currently, or could potentially, offer EFT services—will offer. EFT providers, however, also induce consumers to accept services that are unfamiliar and perhaps not always in the consumer's interest because they must deal with suppliers of the technology, those who actually produce the hardware and software and who supply specific services and equipment. The interests of providers and suppliers usually tend toward promotion and rapid development of EFT technologies.

Standing between these actors and EFT technologies are the government agencies that regulate and control EFT systems, both directly and indirectly, through regulation of the basic interaction among consumers, providers, and suppliers. State and federal regulations not only prescribe the extent and nature of EFT development but often set the framework within which EFT innovations may occur. These private and public institutional actors set the boundaries for the development of EFT technology.

Once developed, EFT operating systems and technologies will have a substantial impact on people and the economy. The impact of EFT may be expected to reinforce or to change any or all of the actors' views of the desirability of further developments. Indeed, evaluation and assessment of the impact of EFT should provide a basis for understanding the dynamics of EFT development and for monitoring and evaluating EFT systems. Given the diverse interests, conflict inevitably will develop among consumers, providers, and suppliers of EFT, and these conflicts will need to be resolved through the policy process.

This chapter focuses on the dynamics of monitoring and evaluating EFT systems as they relate to the impact of EFT on the individual and society. We argue that it is extremely difficult, if not impossible, to establish a monitoring and evaluating system as a structured part of the policy process. The experience of the 1970s indicates that the process of developing EFT technology and the political interactions among the various groups involved have focused mainly on establishing basic ground rules for operating EFT and on assuring that no single financial interest groups receives an unfair competitive edge. The interests of the individual consumer and society as a whole have been largely ignored or only reluctantly addressed. Furthermore, it appears to be extremely difficult to structure an institutional means to monitor the broader, long-range social impact of EFT. Indeed, the problem is that there is not one impact as such; rather, the socioeconomic impact varies depending upon who the actors are and the perspectives they represent. Thus, we question whether it is possible to go beyond an evaluation of the short-term impact of EFT on any single actor or group of actors and to look at the broader, long-run impact on individuals and society.

We begin by setting forth the range of perspectives and values underlying the development of EFT technology and then review the potential impact of EFT, especially for financial institutions. With this background, we examine two efforts to assess the impact of EFT: the completed work of the NCEFT and the current study on EFT by the OTA. Finally, we review the implications of these studies, particularly as they relate to the impact of EFT on the individual consumer and future public-policy research.

Perspectives and Values as They Relate to the Development of EFT Systems

The perspectives and values that an individual brings to studying the dynamics of EFT evolution and development naturally result in different assessments of the impact of EFT and of the parties that should be included in the policymaking process. Five major value orientations, or normative models, implicit in discussions of EFT systems have been identified and examined in the literature[4]:

1. Private-enterprise model: The preeminent consideration of the private-enterprise model is profitability of the EFT systems with the highest social good being the profitability of the firms' providing or utilizing the systems—as long as the competition is fair. Other social goods such as user's privacy or the need for government or data are secondary.
2. Statist model: With the statist model, the strength and efficiency of governmental institutions are the highest goal. Government needs for access to personal data on citizens and for mechanisms to enforce obligations to the state always prevail over other considerations.
3. Libertarian model: With the libertarian model, civil liberties, as specified by the Bill of Rights, are to be maximized in any social choice and in the development of EFT services. Other social services and purposes such as profitability or welfare of the state are to be sacrificed if they conflict with the prerogatives of the individual.
4. Neopopulist model: With the neopopulist model, the practices of public agencies and private enterprises should be easily intelligible to ordinary citizens and responsible to the needs of these individuals. Societal

institutions should emphasize serving the common man and the needs of the individual.

5. Systems model: The main goal with the systems model is that EFT systems be technically well organized, efficient, and reliable. The goal here is on the system as opposed to profitability, role of government institutions, protection of civil liberties, or the needs of the individual.

In spite of these different and rival value orientations, primary support for EFT comes from the private sector, which places a premium on the private-enterprise model. In the last decade, financial institutions have been increasingly concerned about the growing number of checks and resultant check writing as a part of our paper-based financial system. Financial institutions have automated record-keeping systems in order to gain greater efficiency and to provide services to the public that would enable particular financial institutions (with such EFT services) to gain a greater share of consumer business. As EFT services develop, however, they potentially may alter the balance between the various actors in the financial community and thus become a contested part of the financial system.

While Congress has paid significant attention to achieving major reform of financial institutions, the lobbying efforts of financial industries have blocked passage of any legislation. Political pressure for reform was motivated by economic factors in the marketplace and by the development of EFT technology.[5] In fact, while financial institutions experimented with EFT for some time, the real impetus for regulation came only after 1974 when the Federal Home Loan Bank Board gave approval to First Federal Savings and Loan of Lincoln, Nebraska, to place EFT terminals in the Hinky Dinky Supermarkets. With these terminals, customers can make deposits into their savings account, cash checks, and make cash withdrawals from their savings accounts. The impact on financial practices of this technology-based change was substantial, as indicated by Jack Benton, former executive director of the NCEFT:

> This shocked much of the industry. . . . The competitors of First Federal started rushing to install equipment; the Comptroller of the Currency took steps to achieve equality with the S&L's by declaring that EFT terminals were not branches . . .; 21 states rushed to pass EFT legislation . . .; and Congress began to consider legislation establishing a moratorium on EFT developments.[6]

EFT technology not only provided the initial stimulus for change in the financial industry but also made possible certain other kinds of innovations that tend to change the competitive balance among financial institutions. Financial institutions, for example, are not legally allowed to pay interest on checking accounts. However, if a financial institution can automatically transfer funds from a savings account to a checking account through the use of EFT facilities, then the legal distinction is almost mute. In the past, federally chartered savings and loan associations have not been allowed to offer any form of third-party payments or demand-deposit powers. Yet, to the extent that a savings and loan can set up a remote-service unit (RSU) or an automatic teller machine (ATM) in a retail establishment or shopping center, savings and loans acquire a form of third-party payments. Thus, EFT technology provides a means to supercede present legislation and regulation. One further illustration relates to the hotly contested issue of financial-institution branching. The branching of federally chartered banks currently permits legal restrictions adopted by each state. Some states allow statewide branching, while others only permit one branch of any particular bank within its jurisdiction. The question arises, though, as to whether an ATM terminal is a branch. To the extent that an ATM can be used to open an account, withdraw funds, deposit funds, close an account, and so forth, it provides many of the same functions as a brick-and-mortar branch. The courts have held that an ATM has many of the same properties as a brick-and-mortar branch and, therefore, must follow the same legal restrictions. By contrast, the NCEFT argued that a distinction should be drawn between an ATM and a brick-and-mortar branch.[7] While this issue remains unresolved in the public-policy arena, EFT technology clearly is forcing a change in the competitive balance between financial institutions.

In discussing these technology-based changes, our primary concern has not been merely to illustrate the impact of EFT on society and individuals. Rather, our concern lies with how to evaluate the changes and what those changes might bode for the private sector, various competing financial institutions, and the customers of financial institutions.

Efforts to Assess the Individual and Social Impact of EFT Systems

The predominant value orientation, the private-enterprise model, that characterizes the development of EFT technology also tends to epitomize assessments of the individual and social impact of EFT. To illustrate this use of the private enterprise model, we examine two major assessments of EFT: one conducted by the NCEFT and one that is currently being conducted by the OTA.

The NCEFT Study

On 28 October 1974, Congress established the NCEFT. The commission was composed of twenty-six members, with the membership drawn from government agencies, representatives of state regulatory agencies, representatives of financial-institution communities, and public members:

The chairman of the board of governors of the Federal Reserve system or his delegate;

The attorney general or his delegate;

The comptroller of the currency or his delegate;

The chairman of the Federal Home Loan Bank Board or his delegate;

The administrator of the National Credit Union Association or his delegate;

The chairman of the board of directors of the Federal Deposit Insurance Corporation or his delegate;

The chairman of the Federal Communications Commission or his delgate;

The postmaster general or his delegate;

The secretary of the treasury or his delegate;

The chairman of the Federal Trade Commission or his delegate;

Two individuals, appointed by the president, one of whom is an official of a state agency that regulates banking or similar financial institutions, and one of whom is an official of a state agency that regulates thrift or similar financial institutions;

Seven individuals, appointed by the president, who are officers or employees of, or who otherwise represent banking, thrift, or other business entities, including one representative each of commercial banks, mutal savings banks, savings-and-loan associations, credit unions, retailers, nonbanking institutions offering credit-card services, and organizations providing interchange services for credit cards issued by banks;

Five individuals, appointed by the president, from private life, who are not affiliated with, do not represent, and have no substantial interest in any banking, thrift, or other financial institution including, but not limited to, credit unions, retailers, and insurance companies;

The comptroller general of the United States or his delegate;

The director of the OTA.

In addition, Public Law 93–495 indicated that the chairperson shall be designated by the president at the time of his appointment from among the members of the commission and that such selection shall be by and with the advice and consent of the Senate unless the appointee holds an office to which he was appointed by and with the advice and consent of the Senate.

A variety of objectives were assigned to the commission and are presented in the following list:

The need to preserve competition among the financial institutions and other business enterprises using such a system,

The need to promote competition among financial institutions and to assure that government regulation and involvement or participation in a system competitive with the private sector be kept to a minimum,

The need to prevent unfair or discriminatory practices by any financial institution or business enterprise using or desiring to use such a system,

The need to afford maximum user and consumer convenience,

The need to afford maximum user and consumer rights to privacy and confidentiality,

The impact of such a system on economic and monetary policies,

The implications of such a system on the availability of credit,

The implications of such a system expanding internationally and into other forms of electronic communications,

The need to protect the legal rights of users and consumers.

The three main areas of public-policy interest were (1) the need to preserve competition among financial institutions and other business enterprises using such a system, (2) the need to promote competition among financial institutions and to assure that governmental regulations be kept to a minimum, and (3) the need to prevent unfair discriminatory practices by any financial institution or business enterprise using EFT systems. Other considerations included the need for maximum consumer convenience, consumer rights, impact on economic and monetary policy, and the need to protect the legal rights of users and consumers.

In view of the membership of the commission, the first three objectives clearly were destined to become of prime importance. The membership of the NCEFT also indicates that the emphasis and value orientation of the commission would be that of the private-enterprise model. Of the twenty-six members, sixteen were either individuals selected to represent the various members of the financial and computer retail communities (including commercial banks, savings and loans, mutual savings banks, credit unions, retailers, nonbanking institutions offering credit-card services, and organizations providing interchange services for credit cards issued by banks) or regulators of financial institutions at a national or state level. Considering the fact that cabinet heads and other appointed federal administrators were primarily involved with regulation of financial institutions, only five of the twenty-six members of the commission were public members—that is, representatives of consumer interests. Thus, the commission was almost destined to focus on the competitive impact of EFT and to insure that whatever policies developed would accommodate the various interests of financial institutions. Although the impact of EFT on the individual was listed as a factor for consideration, because of the composition and focus of the commission, such concerns inevitably had second priority.[8]

The NCEFT also had a relatively low priority within the executive branch. This is probably best illustrated by the fact that it took over a year for the White House to appoint the commission. The commission operated via a series of acting committees including a users committee, a providers committee, a regulatory-issues committee, and a suppliers committee.

The final report of the commission was released on 28 October 1979. The report summarized the issues covered by each of the four acting committees with major parts dealing with the development of EFT, consumer issues, international developments in EFT, and the roles of the federal government in regulating EFT. The commission made a number of recommendations for EFT but primarily aimed at balancing the competing financial interests and, not surprisingly, the interests of the various members represented on the commission. The thrust of the report was future policy and legislation, not monitoring or assessing the societal impact of EFT.[9]

We would further argue that when Congress established a time framework (two years) for the NCEFT, it insured (consciously or unconsciously) that the policy questions and the answers would be defined largely by the financial and retail community. The short time frame meant that new research could not be conducted and that the commission would rely on information mainly in the hands of financial institutions, their technology suppliers, or their consultants.[10] This emphasis on the present, as opposed to the future, impact of EFT is further stressed in a commentary on the commission by Joseph Coates: "The most important limitation I see on the work to date [of the NCEFT] is the absence of any image of the future. . . . This absence of a vision or framework of the future is the single most critical deficiency in the Commission's work."[11]

The OTA Study

OTA is currently conducting a study of national information systems that focuses on the societal impact of computer technology in three areas: the

NCIC, electronic mail systems (EMS), and EFT. Major differences exist between the studies performed by the NCEFT and the OTA. The OTA study is conducted by a combination of in-house staff and a variety of consultants from around the country with support from advisory panels. The OTA is less interested in sorting out the competitive advantages of these technologies from one industry to another and more interested in a broad-based assessment of the impact of EFT (as well as EMS and the NCIC) on society.

The OTA study nevertheless has a different set of problems that limits its ability to monitor and evaluate the impact of EFT on society and the individual. Most important, the resources for the study are significantly limited—approximately $150,000 for the OTA as compared to $2 million for the NCEFT. This means that resources are simply not available to conduct new research or to evaluate and monitor the impact of EFT on society. Given these financial constraints and, hence, restraints on research, the most that may be expected to be accomplished is the assemblage of past research and discussion of the scenarios for the future of EFT and some of their policy and political implications.

The OTA study has potential, but it remains unclear how much it will contribute to monitoring and evaluating the impact of EFT on society. The OTA advisory group, moreover, has been split between those who are interested in the private-enterprise value orientation and those who lean more toward libertarian or neopopulist value orientations. There is little doubt that each of these factions will strive to insure that their perspective is presented in the OTA study, and thus the final conclusions of the OTA remain uncertain.

The Need for a Broad Agenda to Monitor the Impact of EFT

Jack Benton's post-NCEFT comment that the long run impacts of EFT are largely unknown is likely to also characterize the situation even following the OTA's work. The OTA study is a first attempt to assess the societal impact of EFT, but limited resources and time constraints make it difficult for the OTA to do more than to point to future research and policy agendas.

In order to comprehensively assess the impact of EFT, it is necessary to identify what long-range social changes are likely to occur and to interact with EFT.[12] Changes, for example, in life-style, housing preferences, work patterns, transportation, communications systems, retail and shopping patterns, and similar social forces will affect implementation of EFT. Then, again, EFT itself may be expected to have a distinct impact on the society, given different patterns of social interaction. For instance, what effect will EFT, possibly in conjunction with other transportation and communication networks, have on social mobility, individual households and households at different strata of society, work patterns, and so forth?

We conducted a study and a national conference, sponsored by the NSF, to develop an agenda for EFT-related research.[13] Although the conclusions of that project are reported elsewhere, it is informative to look at the research agenda generated by that study. Table 12-2 outlines the agenda as it relates to EFT impacts on society, the economy, technological issues, regulation and control of EFT, and evaluating and monitoring EFT systems. The agenda provided a bare outline of the kinds of questions that must be addressed prior to evaluating and monitoring the impact of EFT.

Table 12-2
EFT-Research Agenda

Impacts on Society

Impacts of EFT on consumers:
Issue 1.   Costs and benefits of EFT to the consumer
Issue 2.   Consequences of EFT for the less advantaged

Consumer education and protection:
Issue 3.   Education of consumers for dealing with EFT

Record-keeping practices:
Issue 4.   Records control and consumer protection under EFT

Privacy and confidentiality:
Issue 5.   Effective and acceptable privacy safeguards

Economic Impact

Economics of EFT:
Issue 6.   Costs of the current financial-payments system

Economic and financial-institution issues:
Issue 7.   Impact of EFT on the financial system and financial institutions

Technological Issues

Issue 8.   Clarification of technological issues warranting research

Regulation and Control

Need for and targets of regulation:
Issue 9.   Identification and understanding of new regulatory issues arising from EFT
            (including the question of whether regulation is needed at all)
Issue 10. Study of the range and options of organizing EFT-related regulatory structures

Evaluating and Monitoring EFT Systems in the Broader Context

Issue 11. Long-run interactions between EFT and society
Issue 12. EFT as a study of technological change and impact

A further brief discussion of one of these areas, the social impacts of EFT, is especially informational and illustrative of the questions presupposed by a comprehensive assessment of EFT technology.

When people were questioned at the NSF sponsored conference regarding EFT-research priorities, the impact of EFT on people, particularly consumers, received the highest ranking. The costs and benefits of EFT to the consumer was the highest ranked topic both in terms of total points and in terms of highest intensity rating (see tables 12–3 and 12–4).

Table 12–3
Ranking of Research Issues by Participants at the NSF-Sponsored Conference on EFT and Public Policy

Issue	Total Points[a]	Rank among the 37 Issues[a]	Intensity[b]
Technological Issues			
Security	86	15	5.1
Alternative network approaches	84	16	4.9
Alternative communication systems and EFT	82	18	4.8
Reliability	72	23	4.5
Implications of potential	70	24	4.4
Reversibility of EFT changes	68	25	5.7
Impacts on People			
Costs and benefits to the consumer	269	1	7.3
Educating consumers regarding their EFT-related rights	149	5	5.5
Low-income consumers	130	8	5.4
Individual surveillance	95	14	6.3
Consumer abuse (debit cards stolen, payments initiated)	82	19	4.3
Consumer behavior	81	20	4.3
Record controls and counterfeiting	53	28	4.1
Privacy problems	40	33	4.4
Ombudsman as a means of consumer protection	31	35	5.2
Effects of mandatory-disclosure laws on consumer	10	37	3.3
Economic Impact			
Comparative costs of current payment system and EFT	268	2	6.2
Market competition	104	12	4.2
The definition and velocity of money	83	17	4.2
Smaller financial institutions	67	26	3.7
Operating and other expenses	63	27	3.3
EFT and float	45	30	2.8
EFT-induced changes in monetary systems	41	32	4.6
Impact as a result of EFT fraud	22	36	5.5

Table 12–3 continued

Issue	Total Points[a]	Rank among the 37 Issues[a]	Intensity[b]
Regulation and Control			
Definition of EFT regulation: What should be regulated?	202	3	6.1
Access rules for EFT data	143	6	5.7
Federal government as operator of EFT systems	140	7	5.4
Need for EFT regulation	115	11	5.8
Roles of various federal and state legislators/regulators	100	13	5.6
Institutionalizing consumer interests	76	21	4.5
Equal access of all to EFT	73	22	4.6
What private institutions should be regulated (bank, nonbank, and so on)?	51	22	3.6
Evaluating and Monitoring EFT Systems			
Impacts on the long-range character of society	199	4	6.6
Impact on other societal institutions	127	9	5.8
EFT as a case study of technological change and impact	116	10	3.7
EFT development in other countries	44	31	4
EFT and international-fund flows	32	34	4.6

[a]Respondents were asked to rank each of the thirty-seven issues noted above from 1 to 10, with a 10 indicating the item of highest priority, 9 the second high, and so on. The total-points score simply adds up all the points received by an issue.

[b]The intensity score was calculated by dividing the number of times the issue was ranked in the top ten by the total points it received (in other words, if a research issue area has an intensity of 7, it means that, on average, those who ranked this issue in the top ten felt it was the third most important issue.)

Despite the agreement on the importance of consumer issues, disagreement was about how to ensure adequate attention to such issues. Some conference participants felt that market research and competition among financial institutions would ensure that consumer interests are adequately met by EFT developments. However, others were less sanguine about that prospect. The following exchange illustrates this concern:

Coates: As major consultants to the industry, could you tell us what kinds of socially conscious questions come forward from your clients?

Horan: I think that all market research, all planning of banks, of merchants, of depository institutions are premised on the fact that if the con-

Table 12-4
Ranking of Selected EFT Issues according to Institutional Perceptions

Issues	Providers		Regulators		University Researchers		Suppliers		Total
	Financial	Nondepository institutions	State government	Federal government	Business/ economics department	Other university	EFT-technology industry	Other	
Costs and Benefits of EFT to the Consumer									
Number of people responding	24	1	2	17	2	9	6	11	72
Number of times ranked in the top ten	14	1	2	5	2	6	2	5	37
Total points	103	10	17	41	19	26	14	39	269
Intensity	7.4	10	8.5	8.2	9.5	4.3	7	7.8	7
Comparative Costs of the Current Payments System									
Number of people responding	24	1	2	17	2	9	6	11	72
Number of times ranked in the top ten	19	1	2	8	2	3	4	4	43
Total points	119	8	12	42	19	14	25	29	268
Intensity	6.2	8	6	5.3	9.5	4.7	6.3	7.3	6

Definition of EFT Regulation:
What should be regulated?

Number of people responding	24	1	2	17	2	9	6	11	72
Number of times ranked in the top ten	11	0	1	10	1	4	3	2	33
Total points	64	0	3	79	7	4	24	21	202
Intensity	5.8	0	3	7.9	7	1	8	7	6

EFT Impact on the Long-Range
Character of Society

Number of people responding	24	1	2	17	2	9	16	11	72
Number of times ranked in the top ten	9	0	2	4	0	7	2	6	30
Total points	40	0	15	39	0	54	9	42	199
Intensity	4.4	0	7.5	9.8	0	7.7	4.5	7	6

EFT as a Case Study of Techno-
logical Change and Impact

Number of people responding	24	1	2	17	2	9	6	11	72
Number of times ranked in the top ten	7	1	0	7	2	7	3	4	31
Total points	14	6	0	11	9	46	13	17	116
Intensity	2	6	0	1.6	4.5	6.6	4.3	4.3	3

sumer doesn't accept it, you just don't have a market. It's a commercial type of decision.

Coates: I take it the answer is none.

Louderback: I see my clients thinking very carefully about the services that they are going to be offering their customers, their depositors, thinking very carefully because if they offer the right kind of services and satisfy the right kinds of needs, they are going to be more successful than their competitors.

Coates: [I]f one says that competition will protect the consumer, I might be willing to accept that in the competitive EFT market, such as POS, point of sale, is today. But I don't think it's an adequate response in a more concentrated market, such as with automated clearinghouses, where for preauthorized debits and credits there is only one game in town, through the association.[14]

In addition to illustrating differences in perspectives on EFT research, this exchange indicates that EFT impact on consumers might vary considerably depending upon the specific EFT technology implemented. For example, ATMs seem to have been well received by consumers, whereas many POS terminals appear to have been poorly received by both consumers and retailers.[15]

On the basis of our preliminary study, four main research issues apparently need to be explored if we are to assess the social impact of EFT systems.

The first issue is the costs and benefits of EFT to the consumer. Little is known about how much EFT will cost and how those costs will be passed on to consumers. Consumers nonetheless will eventually bear most of the costs regardless of whether they pay through transaction charges or taxes for governmental subsidy of EFT or both. Similarly, little is known about what benefits EFT will bring to consumers and how those benefits will be distributed. Projected benefits include reduced costs of payment services, increased consumer convenience, increased security of financial transactions, and the like. However, the extent to which these intended consequences actually occur remains uncertain. Unintended and unanticipated consequences are even less clear. Whether the purported benefits of EFT are desired by consumers is even unknown.[16] More important, which group of consumers will receive the benefits and which will pay the costs of EFT remains problematic. Some indications are that high-income, well-educated, financially sophisticated, credit-card-using consumers are the most likely users of EFT services.[17] Yet, if costs are allocated generally over the public and if benefits accrue disproportionately to some minority of consumers, the potential exists for a kind of unfair taxation.

The second issue concerns the consequences of EFT for the less-advantaged members of society. A large group of people is likely to be excluded from EFT services because it is comprised of disadvantaged, "unbanked," people who constitute 25 percent or more of the population.[18] Without governmental action, the net impact of EFT is likely to be negative: "At the very least, policymakers should take care that this new type of financial institution does not promote more inequality."[19] More than any other, the equity issue illustrates the potential conflict between the values of the private-enterprise perspective and the more socially conscious neopopulist perspective.

The third issue requiring research concerns the education of consumers who deal with EFT. Any effort to stimulate widespread adoption of EFT systems would seem to require a coordinated educational program of considerable scale.[20] It is critical that consumers are able to make reasonable choices about whether to adopt the EFT services. Studies are needed to determine the kinds of educational programs necessary to develop public awareness of the EFT and, assuming EFT develops as anticipated, to facilitate informed and responsible consumer use of the system.

The fourth issue concerns records control and consumer protection under EFT. Control over financial records and prevention of fraud and other abuses are potential problems posed by EFT.[21] As Kathleen O'Reilly observed at our conference on EFT technology:

> Consumers are becoming terribly concerned about the implications of the computer fraud phenomenon. It is far from science fiction. If there is not a commitment to the development of EFT systems that guarantee that appropriate (and available) technological methodology is used to minimize computer fraud (the prime victims of which are consumers), EFTs may well enhance the opportunity for that kind of dangerous abuse.[22]

The changes posed by EFT with regard to the philosophy and procedures of financial record keeping and consumer protection might indeed be dramatic. Indeed, one cricital change involves the very definition of what constitutes money. EFT represents a major move away from the use of cash to back up the symbols representing assets and liabilities. In such an event, what will be the standard for accounting for a given amount of money? An electronic impulse of a certain characteristic over an authorized channel? EFT presents wide-ranging implications for accounting and auditing practices, as well as for protection of financial records.

Conclusion

Only a few years ago, many technological and financial experts predicted that EFT systems would usher in the checkless/cashless society. That has not happened, though the potential still exists for major change. Incorporation and routinization of EFT technology has occurred much more slowly than the early technologists and promoters expected. [23] Most significantly, EFT technology has been shaped by the organizational and institutional context in which it is used rather than serving as a driving force itself. EFT technology appears to be shaped more by the agendas of those who use it than by the possibilities inherent in the technology. [24] EFT services, therefore, may be expected to exhibit an evolution—that is, many small incremental changes and adaptations to changing organizational and societal definitions of the appropriate use of EFT systems.

From a public-policy perspective, the issues surrounding EFT technology will remain on the public agenda for years to come; and EFT policy and technology will be mixed. As EFT technology changes and is adapted to new uses, public policy will evolve to deal with the social impact of the technology. As we develop greater understanding of the ways in which the society might utilize EFT technology, public policy may also be directed increasingly toward shaping the way it is utilized.

From the standpoint of research, the joint evolution of EFT technology and the public-policy agenda means that a continuous monitoring and evaluation of EFT is paramount. Objective, scientific information can go a long way toward informing public policymaking. However, we have argued that it is extremely difficult, if not impossible, to establish mechanisms to evaluate and monitor the impact of EFT systems as part of the policy process. The time frame for policymaking has been too short to facilitate needed research and the value orientations of the participants in the policy process too one sided to produce balanced research.

We therefore suggest that EFT research be conducted outside the policy process. One approach, advocated by John King and Kenneth Kraemer, is to create a national center for the continuous study of EFT in much the same manner as we currently study the weather. [25] They advocated that the center conduct research on several broad areas of the ineraction of technology and society—for example, how technology emerges, how it is handled by existing institutions, what specific impacts new technologies have, and how technologies change over time to conform to new circumstances and developments. Each of these research areas for assessing EFT technology has academic, public-policy, and practical relevance.

Concentration of technology-related research in a single center, no matter how desirable technically, nevertheless might result in the same political problems that have plagued past EFT studies at the national level. Consequently, we suggest that technology-assessment research should be decentralized in multiple centers—mainly in universities but possibly also in some federal agencies. Such centers would require five-to-ten-year charters, funded on a continuous basis by an institution like the NSF and advised by representatives of the respective interests in EFT technology. The center would offer an opportunity to produce continuous high-quality research from a core staff committed to EFT research as a policy, intellectual, and practical focus. Such a center would generate a solid information base for policy decisions and reduce the need for ad hoc approaches to this critical technological development.

Notes

1. National Commission on Electronic Funds Transfer, *Final Report on EFT in the United States: Policy Recommendations and the Public Interest* (Washington, D.C.: Government Printing Office, 28 October 1977).

2. For a discussion of such developments, see, for example, Kent W. Colton and Kenneth L. Kraemer, eds., *Computers and Banking, Electronic Funds Transfer Systems and Public Policy* (New York: Plenum Publishing Co., 1980); Arthur D. Little, Inc., *The Consequences of Electronic Funds Transfer: A Technology Assessment of Movement toward a Less Cash/Less Check Society,* Rept. C76397, National Science Foundation (Cambridge, Mass., January 1975); K.H. Humes, "The Checkless/Cashless Society? Don't Bank on It!" *The Futurist* (October 1978):301–306; National Commission on Electronic Funds Transfers, *EFT in the United States;* and Peat, Marwick and Mitchell and Co., *EFT: A Strategy Perspective* (New York, 1977).

3. A more-complete discussion of this conceptual framework is contained in Colton and Kraemer, *Computers and Banking,* chapter 1.

4. This discussion draws on Rob Kling, "Value Conflicts and Social Choice in Electronic Funds Transfer System Developments," *Communications of the ACM* 21 (August 1978):642–647.

5. For a further discussion of the process of change related to financial institutions and the development of EFT, see Kent W. Colton, "Financial Reform: A Review of the Past and Prospects for the Future," *Journal of the American Real Estate and Urban Economics Association* (Summer 1980):43.

6. See Imperial Computer Sciences, Inc., "Electronic Funds Transfer: The Policy Issues," Working Paper (Washington, D.C.: U.S. Congress, Office of Technology Assessment, Telecommunications and Information Systems Group, May 1979), p. 14.

7. National Commission on Electronic Funds Transfers, *EFT in the United States.*

8. In fact, critics have argued that the NCEFT was simply an extension of the special interests lobbying in Congress and that the commission was established as a means of putting off congressional action for a two-year period.

9. See, for example, Imperial Computer Sciences, "Electronic Funds Transfer," p. 15.

10. W. Boucher, Working Paper no. PPR7704 "A Comment on EFT Research: Five Tries and a Start," (Cambridge, Mass.: Public Systems Evaluation, Inc., 1976).

11. Found in Kenneth L. Kraemer and Kent W. Colton, "An Agenda for EFT Research," in *Computers and Banking, Electronic Funds Transfer Systems and Public Policy,* Colton and Kraemer, eds., chapter 19 (New York: Plenum Publishing Co., 1980).

12. See, for example, ibid., chapter 14; and R. Kling, "EFT and the Quality of Life," in *Proceedings of the NCC,* vol. 47 (Alexandria, Va.: AFIPS Press, 1978), pp. 191–197.

13. See Kenneth L. Kraemer and Kent W. Colton, "Policy Values and Research: Anatomy of a Research Agenda," *Communications of the ACM* (December 1979):175.

14. Colton and Kraemer, *Computers and Banking,* chapter 19.

15. For a further discussion, see "A Retreat from the Cashless Society," *Business Week,* 18 April 1977, pp. 80–90; J.F. Fisher, "EFT— The Decade of the 1980's: New Concepts for the World of Banking," *The Banker's Magazine* 162 (March/April 1979):2; and S. Pastore, "EFT and the Consumer," *The Banker's Magazine* 162 (March/April 1979):35–42.

16. See "Retreat from the Cashless Society."

17. See Humes, "Checkless/Cashless Society?"

18. See S.R. Hiltz and M. Turoff, "EFT and Social Stratification in the U.S.A.: More Inequality?" *Telecommunications Policy* 2 (March 1978):22–31.

19. Ibid., pp. 221–231.

20. For a further discussion, see R. Kling, "Passing the Digital Buck: Unresolved Social and Technical Issues in Electronic Funds Transfer Systems," (Irvine: University of California, Information and Computer Science Department, June 1976), Paper no. TR 87; J. Rule, "Value Choices in Electronic Fund Transfer Policy," (Washington, D.C.: Office of Telecommunications Policy, Executive Office of the President, October 1975); and K. Sayre, ed., *Values in the Electronics Power Industry* (Notre Dame, In.: University of Notre Dame Press, 1977).

21. For a further discussion, see Kling, "Passing the Digital Buck," D.B. Parker, *Crime by Computer* (New York: Scribner, 1976); P.S. Portway, "EFT Systems? No Thanks, Not Yet," *Computerworld* 12 (9 January 1978):14–16, 21, 23–25; Privacy Protection Study Commission, *Personal Privacy in an Information Society* (Washington, D.C.: Government Printing Office, July 1977); L.W. Rossman, "Financial Industry Sees EFT Privacy Laws Adequate," *American Banker* 210 (28 October 1976):2, 11; J. Rule, *Private Lives and Public Surveillance* (New York: Schocken Books,

1974); Rule, "Value Choices"; and A. Westin and M. Baker, *Databanks in a Free Society* (New York: Quadrangle Books, 1972).

22. F. Leary, Jr., K. O'Reilly, and H.M. Palmer, "Discussion of Papers by K. Humes, S.R. Hiltz and M. Turoff," Working Paper no. PPR-7709 (Cambridge, Mass.: Public Systems Evaluation, Inc., 1976).

23. See, for example, L. Winner, *Autonomous Technology: Technology Out of Control as a Theme in Political Thought* (Cambridge, Mass.: M.I.T. Press, 1977).

24. See John Leslie King and Kenneth L. Kraemer, "EFT as a Subject of Study in Technology, Society, and Public Policy," *Telecommunications Policy* 2 (March 1978):13–21.

25. Ibid.

Appendix:
The Technology
Assessment Act of 1972

(Pub. L. 92–484, §2, 13 October 1972, 86 Stat. 797, 42 U.S.C. 1862)

§471. Congressional findings and declaration of purpose

The Congress hereby finds and declares that
(a) As technology continues to change and expand rapidly, its applications are
    (1) large and growing in scale; and
    (2) increasingly extensive, pervasive, and critical in their impact, beneficial and adverse, on the natural and social environment.

(b) Therefore, it is essential that, to the fullest extent possible, the consequences of technological applications be anticipated, understood, and considered in determination of public policy on existing and emerging national problems.

(c) The Congress further finds that
    (1) the Federal agencies presently responsible directly to the Congress are not designed to provide the legislative branch with adequate and timely information, independently developed, relating to the potential impact of technological applications, and
    (2) the present mechanisms of the Congress do not and are not designed to provide the legislative branch with such information.

(d) Accordingly, it is necessary for the Congress to
    (1) equip itself with new and effective means for securing competent, unbiased information concerning the physical, biological, economic, social, and political effects of such applications; and
    (2) utilize this information, whenever appropriate, as one factor in the legislative assessment of matters pending before the Congress, particularly in those instances where the Federal Government may be called upon to consider support for, or management or regulation of, technological applications.

(Pub. L. 92–484, §2, 13 October 13 1972, 86 Stat. 797)

Short Title

Section 1 of Pub. L. 92–484 provided: "That this Act which enacted this chapter and amended section 1862 of Title 42, The Public Health and Welfare, may be cited as the 'Technology Assessment Act of 1972'."

Section Referred to in Other Sections

This section is referred to in section 472 of this title.

§472. Office of Technology Assessment

(a) Creation

In accordance with the findings and declaration of purpose in section 471 of this title, there is hereby created the Office of Technology Assessment (hereinafter referred to as the "Office") which shall be within and responsible to the legislative branch of the Government.

(b) Composition

The Office shall consist of a Technology Assessment Board (hereinafter referred to as the "Board") which shall formulate and promulgate the policies of the Office, and a Director who shall carry out such policies and administer the operations of the Office.

(c) Functions and duties

The basic function of the Office shall be to provide early indications of the probable beneficial and adverse impacts of the applications of technology and to develop others to coordinate information which may assist the Congress. In carrying out such function, the Office shall

(1) identify existing or probable impacts of technology or technological programs;

(2) where possible, ascertain cause-and-effect relationships;

(3) identify alternative technological methods of implementing specific programs;

(4) identify alternative programs for achieving requisite goals;

(5) make estimates and comparisons of the impacts of alternative methods and programs;

(6) present findings of completed analyses to the appropriate legislative authorities;

(7) identify areas where additional research or data collection is required to provide adequate support for the assessments and estimates described in paragraph (1) through (5) of this subsection; and

(8) undertake such additional associated activities as the appropriate authorities specified under subsection (d) of this section may direct.

(d) Initiation of assessment activities
Assessment activities undertaken by the Office may be initiated upon the request of
(1) the chairman of any standing, special, or select committee of either House of the Congress, or of any joint committee of the Congress acting for himself or at the request of the ranking minority member or a majority of the committee members;
(2) the Board; or
(3) the Director, in consultation with the Board.

(e) Availability of information
Assessments made by the Office, including information, surveys, studies, reports, and findings related thereto, shall be made available to the initiating committee or other appropriate committees of the Congress. In addition, any such information, surveys, studies, reports, and findings produced by the Office may be made available to the public except where
(1) to do so would violate security statutes; or
(2) the Board considers it necessary or advisable to withhold such information in accordance with one or more of the numbered paragraphs in section 552(b) of title 6.

(Pub.L. 92–484, §3, 13 October 1972, 86 Stat. 797)

Section Referred to in Other Sections

This section is referred to in section 476 of this title.

§473. Technology Assessment Board

(a) Membership
The Board shall consist of thirteen members as follows:
(1) six Members of the Senate, appointed by the President pro tempore of the Senate, three from the majority party and three from the minority party;
(2) six Members of the House of Representatives appointed by the Speaker of the House of Representatives, three from the majority party and three from the minority party; and
(3) the Director, who shall not be a voting member.

(b) Execution of functions during vacancies; filling of vacancies
Vacancies in the membership of the Board shall not affect the power of the remaining members to execute the functions of the Board and shall be filled in the same manner as in the case of the original appointment.

(c) Chairman and vice chairman, selection procedure
The Board shall select a chairman and a vice chairman from among its members at the beginning of each Congress. The vice chairman shall act in the place and stead of the chairman in the absence of the chairman. The chairmanship and the vice chairmanship shall alternate between the Senate and the House of Representatives with each Congress. The chairman during each even-numbered Congress shall be selected by the Members of the House of Representatives on the Board from among their number. The vice chairman during each Congress shall be chosen in the same manner from that House of Congress of which the chairman is a Member.

(d) Meetings; powers of Board
The Board is authorized to sit and act at such places and times during the sessions, recesses, and adjourned periods of Congress, and upon a vote of a majority of its members, to require by subpoena or otherwise the attendance of such witnesses and the production of such books, papers, and documents, to administer such oaths and affirmations, to take such testimony, to procure such printing and binding, and to make such expenditures, as it deems advisable. The Board may make such rules respecting its organization and procedures as it deems necessary, except that no recommendation shall be reported from the Board unless a majority of the Board assent. Subpoenas may be issued over the signature of the chairman of the Board or any voting member designated by him or by the Board, and may be served by such person or persons as may be designated by such chairman or member. The chairman of the Board or any voting member thereof may administer oaths or affirmations to witnesses.

(Pub. L. 92–484, §4, 13 October 1972, 86 Stat. 798)

§474. Director of Office of Technology Assessment

(a) Appointment; term; compensation
The Dirctor of the Office of Technology Assessment shall be appointed by the Board and shall serve for a term of six years unless sooner removed by the Board. He shall receive basic pay at the rate provided for level III of the Executive Schedule under section 5314 of title 5.

(b) Powers and duties
    In addition to the powers and duties vested in him by this chapter, the
    Director shall exercise such powers and duties as may be delegated to
    him by the Board.

(c) Deputy Director; appointment, functions; compensation
    The Director may appoint with the approval of the Board, a Deputy
    Director who shall perform such functions as the Director may pre-
    scribe and who shall be Acting Director during the absence or incapac-
    ity of the Director or in the event of a vacancy in the office of Director.
    The Deputy Director shall receive basic pay at the rate provided for
    level IV of the Executive Schedule under section 5315 of title 5.

(d) Restrictions on outside employment activities of Director and Deputy
    Director
    Neither the Director nor the Deputy Director shall engage in any other
    business, vocation, or employment than that of serving as such Director
    or Deputy Director, as the case may be; nor shall the Director or Dep-
    uty Director, except with the approval of the Board, hold any office in,
    or act in any capacity for, any organization, agency, or institution with
    which the Office makes any contact or other arrangement under this
    chapter.

Pub. L. 92–484, §5, 13 October 1972, 86 Stat. 799)

References in Text

"This chapter," referred to in subsecs. (b) and (d), was in the original "this
Act," meaning Pub. L. 92–484, which enacted this chapter and amended
section 1862 of Title 42, The Public Health and Welfare.

§475. Powers of Office of Technology Assessment

(a) Use of public and private personnel and organizations; formation of
    special ad hoc task forces; contracts with governmental, etc., agencies
    and instrumentalities; advance, progress, and other payments; utiliza-
    tion of services of volunatry and uncompensated personnel; acquisi-
    tion, holding, and disposal of real and personal property; promulgation
    of rules and regulations.
The office shall have the authority, within the limits of available appropriations, to do all things necessary to carry out the provisions of this chapter,
including, but without being limited to, the authority to

(1) make full use of competent personnel and organizations outside the Office, public or private, and form special ad hoc task forces or make other arrangements when appropriate;

(2) enter into contracts or other arrangements as may be necessary for the conduct of the work of the Office with any agency or instrumentality of the United States, with any State, territory, or possession or any political subdivision thereof, or with any person, firm, association, corporation, or educational institution, with or without reimbursement, without performance or other bonds, and without regard to section 5 of title 41;

(3) make advance, progress, and other payments which relate to technology assessment without regard to the provisions of sections 529 of title 31;

(4) accept and utilize the services of voluntary and uncompensated personnel necessary for the conduct of the work of the Office and provide transportation and subsistence as authorized by section 5703 of title 5, for persons serving without compensation;

(5) acquire by purchase, lease, loan, or gift, and hold and dispose of by sale, lease, or loan, real and personal property of all kinds necessary for or resulting from the exercise of authority granted by this chapter; and

(6) prescribe such rules and regulations as it deems necessary governing the operation and organization of the Office.

(b) Recordkeeping by contractors and other parties entering into contracts and other arrangements with Office; availability of books and records to Office and Comptroller General for audit and examination

Contractors and other parties entering into contracts and other arrangements under this section which involve costs to the Government shall maintain such books and related records as will facilitate an effective audit in such detail and in such manner as shall be prescribed by the Office, and such books and records (and related documents and papers) shall be available to the Office and the Comptroller General of the United States, or any of their duly authorized representatives, for the purpose of audit and examination.

(c) Operation of laboratories, pilot plants, or test facilities

The Office, in carrying out the provisions of this chapter, shall not, itself, operate any laboratories, pilot plants, or test facilities.

(d) Requests to executive departments or agencies for information, suggestions, estimates, statistics, and technical assistance; duty of executive departments and agencies to furnish information, etc.

The Office is authorized to secure directly from any executive department or agency information, suggestions, estimates, statistics, and technical assistance for the purpose of carrying out its functions under this chapter. Each such executive department or agency shall furnish the information, suggestions, estimates, statistics, and technical assistance directly to the Office upon its request.

(e) Requests to heads of executive departments or agencies for detail of personnel; reimbursement
On request of the Office, the head of any executive department or agency may detail, with or without reimbursement, any of its personnel to assist the Office in carrying out its functions under this chapter.

(f) Appointment and compensation of personnel
The Director shall, in accordance with such policies as the Board shall prescribe, appoint and fix the compensation of such personnel as may be necessary to carry out the provisions of this chapter.

(Pub. L. 92–484 §6, 13 October 1972, 86 Stat. 799)

References in Text

"This chapter," referred to in text, was in the original "this Act," meaning Pub. L. 92–484, which enacted this chapter and amended section 1862 of Title 42, The Public Health and Welfare.

§476. Technology Assessment Advisory Council

(a) Establishment; composition
The Office shall establish a Technology Assessment Advisory Council (hereinafter referred to as the "Council"). The Council shall be composed of the following twelve members:
(1) ten members from the public, to be appointed by the Board, who shall be persons eminent in one or more fields of the physical, biological, or social sciences or engineering or experienced in the administration of technological activities, or who may be judged qualified on the basis of contributions made to educational or public activities;
(2) the Comptroller General; and
(3) the Director of the Congressional Research Service of the Library of Congress.

(b) Duties
The Council, upon request by the Board, shall
(1) review and make recommendations to the Board on activities under-

taken by the Office or on the initiation thereof in accordance with section 472(d) of this title;

(2) review and make recommendations to the Board on the findings of any assessment made by or for the Office; and

(3) undertake such additional related tasks as the Board may direct.

(c) Chairman and Vice Chairman; election by Council from members appointed from public; terms and conditions of service
The Council by majority vote, shall elect from its members appointed under subsection (a)(1) of this section a Chairman and a Vice Chairman, who shall serve for such time and under such conditions as the Council may prescribe. In the absence of the Chairman, or in the event of his incapacity, the Vice Chairman shall act as Chairman.

(d) Terms of office of members appointed from public; reappointment
The term of office of each member of the Council appointed under subsection (a)(1) of this section shall be four years except that any such member appointed to fill the vacancy occurring prior to the expiration of the term for which his predecessor was appointed shall be appointed for the remainder of such term. No person shall be appointed a member of the Council under subsection (a)(1) of this section more than twice. Terms of the members appointed under subsection (a)(1) of this section shall be staggered so as to establish a rotating membership according to such method as the Board may devise.

(e) Payment to Comptroller General and Director of Congressional Research Service to travel and other necessary expenses; payment to members appointed from public of compensation and reimbursement for travel, subsistence, and other necessary expenses
(1) The members of the Council other than those appointed under subsection (a)(1) of this section shall receive no pay for their services as members of the Council, but shall be allowed necessary travel expenses (or, in the alternative mileage for use of privately owned vehicles and a per diem in lieu of subsistence at not to exceed the rate prescribed in sections 5702 and 5704 of title 5), and other necessary expenses incurred by them in the performance of duties vested in the Council, without regard to the provisions of subchapter 1 of chapter 57 and section 5731 of title 5, and regulations promulgated thereunder.

(2) The members of the Council appointed under subsection (a)(1) of this section shall receive compensation for each day engaged in the actual performance of duties vested in the Council at rates of pay not in excess of the daily equivalent of the highest rate of basic pay

set forth in the General Schedule of section 5332(a) of title 5, and in addition shall be reimbursed for travel, subsistence, and other necessary expenses in the manner provided for other members of the Council under paragraph (1) of this subsection.

Pub. L. 92–484, §7, 13 October 1972, 86 Stat. 800)

Termination of Advisory Councils

Advisory Councils in existence on Jan. 5, 1973, to terminate not later than the expiration of the two-year period following Jan. 5, 1973, unless, in the case of a council established by the President or an officer of the Federal Government, such council is renewed by appropriate action prior to the expiration of such two-year period, or in the case of a council established by the Congress, its duration is otherwise provided by law, see sections 3(2) and 14 of Pub. L. 92–463, Oct. 6, 1972, 86 Stat. 770, set out in the Appendix to Title 5, Government Organization and Employees.

§477. Utilization of services of Library of Congress

(a) Authority of Librarian to make available services and assistance of Congressional Research Service
To carry out the objectives of this chapter, the Librarian of Congress is authorized to make available to the Office such services and assistance of the Congressional Research Service as may be appropriate and feasible.

(b) Scope of services and assistance
Such services and assistance made available to the Office shall include, but not be limited to, all of the services and assistance which the Congressional Research Service is otherwise authorized to provide to the Congress.

(c) Services or responsibilities performed by Congressional Research Service for Congress not altered or modified; authority of Librarian to establish within Congressional Research Service additional divisions, etc.
Nothing in this section shall alter or modify any services or responsibilities, other than those performed for the Office, which the Congressional Research Service under law performs for or on behalf of the Congress. The Librarian is, however, authorized to establish within the Congressional

Research Service such additional divisions, groups, or other organizational entities as may be necessary to carry out the purpose of this chapter.

(d) Reimbursement for services and assistance
    Services and assistance made available to the Office by the Congressional Research Service in accordance with this section may be provided with or without reimbursement from funds of the Office, as agreed upon by the Board and the Librarian of Congress.

(Pub. L. 92–484, §8, 13 October 1972, 86 Stat. 801)

References in Text

"This chapter," referred to in subsecs. (a) and (c), was in the original "this Act," meaning Pub. L. 92–484, which enacted this chapter and amended section 1862 of Title 42, The Public Health and Welfare.

§478. Utilization of services of General Accounting Offices

(a) Authority of General Accounting Office to furnish financial and administrative services.
    Financial and administrative services (including those related to budgeting, accounting, financial reporting, personnel, and procurement) and such other services as may be appropriate shall be provided the Office by the General Accounting Office.

(b) Scope of services and assistance
    Such services and assistance to the Office shall include, but not be limited to, all of the services and assistance which the General Accounting Office is otherwise authorized to provide to the Congress.

(c) Services or responsibilities performed by General Accounting Office for Congress not altered or modified
    Nothing in this section shall alter or modify any services or responsibilities, other than those performed for the Office, which the General Accounting Office under law performs for or on behalf of the Congress.

(d) Reimbursement for services and assistance
    Services and assistance made available to the Office by the General Accounting Office in accordance with this section may be provided

with or without reimbursement from funds of the Office, as agreed upon by the Board and the Comptroller General.

(Pub. L. 92–484, §9, 13 October 1972, 86 Stat. 802)

§479. Coordination of activities with National Science Foundation

The Office shall maintain a continuing liaison with the National Science Foundation with respect to
  (1) grants and contracts formulated or activated by the Foundation which are for purposes of technology assessment; and
  (2) the promotion of coordination in areas of technology assessment, and the avoidance of unnecessary duplication or overlapping of research activities in the development of technology assessment techniques and programs.

(Pub. L. 92–484, §10(a), 13 October 1972, 86 Stat. 802)

§480. Annual report to Congress

The Office shall submit to the Congress an annual report which shall include, but not be limited to, an evaluation of technology assessment technique and identification, insofar as may be feasible, of technological areas and programs requiring future analysis. Such report shall be submitted not later than March 15 of each year.

(Pub. L. 92–484, §11, 13 October, 1972, 86 Stat. 802)

§481. Authorization of appropriations; availability of appropriations

(a) To enable the Office to carry out its powers and duties, there is hereby authorized to be appropriated to the Office, out of any money in the Treasury not otherwise appropriated, not to exceed $5,000,000 in the aggregate for the two fiscal years ending June 30, 1973, and June 30, 1974, and thereafter such sums as may be necessary.

(b) Appropriations made pursuant to the authority provided in subsection (a) of this section shall remain available for obligation, for expenditure, or for obligation and expenditure for such period or periods as may be specified in the Act making such appropriations.

Bibliography

Abelson, Philip H. "The Tris Controversy." *Science* 197 (8 July 1977):113.

Ackerman, Bruce A., and William T. Hassler. "Beyond the New Deal: Coal and the Clean Air Act." *Yale Law Journal* 89 (1980):1466.

————. *Clean Coal/Dirty Air.* New Haven: Yale University Press, 1981.

Ackerman, Bruce A.; Susan Rose-Ackerman; James Sawyer; and Dale Henderson. *The Uncertain Search for Environmental Quality.* New York: Free Press, 1974.

Ackoff, Russell. *Redesigning the Future: A Systems Approach to Society.* New York: John Wiley, 1974.

American Bar Association. *Federal Regulation: Roads to Reform.* Washington, D.C., 1979.

American Cancer Society. *Cancer Facts and Figures: 1981.* New York, 1980.

Ames, Bruce. *"Identifying Environmental Chemicals Causing Mutations and Cancer." Science* 204 (1979):587.

Ames, Charles, and Steven McCracken. "Framing Regulatory Standards to Avoid Formal Adjudication: The FDA as a Case Study." *California Law Review* 64 (1976):14.

Anderson, Frederick R., et al., *Environmental Improvements through Economic Incentives.* Baltimore: Johns Hopkins University Press, 1977.

Anderson, James E. *Public Policy-Making.* New York: Praeger, 1975.

"A Retreat from the Cashless Society." *Business Week,* 18 April 1977, p.80.

Argyris, Chris. *Management and Organizational Development.* New York: McGraw-Hill, 1970.

Armstrong, Joe E., with Willis W. Harman. *Strategies for Conducting Technology Assessments.* Boulder: Westview Press, 1980.

Arnstein, S.R., and A. Christakis. *Perspectives on Technology Assessment.* Jerusalem: Science and Technology Publishers, 1975.

Arthur D. Little, Inc. *The Consequences of Electronic Funds Transfer: A Technology Assessment of Movement toward a Less Cash/Less Check Society.* Cambridge, Mass., 1975.

Ashford, Nicholas A., and G.R. Heaton. "The Effects of Health and Environmental Safety Regulation on Technological Change in the Chemical Industry." In *Federal Regulation and Chemical Innovation,* edited by C.T. Hill. Washington, D.C.: American Chemical Society, 1979, pp. 45–67.

Averbach, Carl A. "Informal Rule Making." *Northwestern University Law Review* 72 (1977):15.

Ayres, Robert, and Allen Kneese. "Production, Consumption and Externalities," *American Economic Review* 59 (June 1969):282.

Baram, Michael S. "Social Control of Science and Technology." *Science* 172 (1971):535.

————. *Regulation of Health, Safety and Environmental Quality and the Use of Cost-Benefit Analysis.* Washington, D.C.: Administrative Conference of the U.S., 1979.

————. "Cost-Benefit Analysis: An Inadequate Basis for Health, Safety and Environmental Regulatory Decisionmaking." *Ecology Law Quarterly* 8 (1980):473.

————. *Alternatives to Regulation.* Lexington, Mass.: Lexington Books, D.C. Heath and Company, 1982.

Baram, Michael S., and J. Raymond Miyares. "The Legal Framework for Determining Unreasonable Risk from Carcinogenic Chemicals." Unpublished report for the Office of Technology Assessment, Washington, D.C., 1980.

Bartlett, Robert V. *The Reserve Mining Controversy: A Case Study of Science, Technology, and Values.* Bloomington: Indiana University, Advanced Studies in Science, Technology, and Public Policy, 1979.

Bauer, Raymond, A., and K.J. Gergen, eds. *The Study of Policy Formation.* New York: Free Press, 1968.

Baxter, William F. "The SST: From Harlem to Watts in Two Hours." *Stanford Law Review* 21 (1968):1.

Bazelon, David. "Coping with Technology through the Legal Process." *Cornell Law Review* 62 (1977):817.

Bell, Daniel. *The Coming of Post-Industrial Society.* New York: Basic Books, 1973.

Bennis, Warren, and Phillip Slater. *The Temporary Society.* New York: Harper & Row, 1968.

Benveniste, Guy. *Regulation and Planning: The Case of Environmental Politics.* San Francisco: Boyd and Fraser, 1981.

Berg, Mark, et al., *Factors Affecting Utilization of Technology Assessment Studies in Policy-Making.* Ann Arbor, Mich.: Institute of Social Research, 1978.

Berg, Paul, et al., "Summary Statement of the Asilomar Conference on Recombinant DNA Molecules." *Proceedings of the National Academy of Sciences,* p. 72. Washington, D.C.: National Academy of Sciences, June 1975.

Berry, Jeffrey, M. *Lobbying for the People.* Princeton, N.J.: Princeton University Press, 1977.

"Biotechnology Becomes a Goldrush." *The Economist,* 13 June 1981, p. 81.

Birnbaum, Sheila. *Toxic Substances: Problems in Litigation.* New York: Practicing Law Institute, 1981.

Blank, Charles H. "The Delaney Clause: Technical Naivete and Scientific Advocacy in the Formulation of Public Health Policies." *California Law Review* 62 (1974):1084.

Blank, Robert H. *The Political Implications of Human Genetic Technology.* Boulder: Westview Press, 1981.

Boeckmann, Margaret. "Policy Impacts of the New Jersey Income Maintenance Experiment." *Policy Sciences* 7 (1976):53.

Boorstin, Daniel J. *The Republic of Technology.* New York: Harper & Row, 1978.

Boulding, Kenneth. *Ecodynamics.* Beverly Hills, Calif. Sage, 1979.

Boyer, Barry. "Alternatives to Administrative Trial-Type Hearings to Resolve Complex Scientific, Economic, and Social Issues. *Michigan Law Review* 71 (1972):iii.

Breyer, Stephen. "Vermont Yankee and the Courts' Role in the Nuclear Energy Controversy." *Harvard Law Review* 91 (1978):1833.

———. "Analyzing Regulatory Failure: Mismatches, Less Restrictive Alternatives, and Reform." *Harvard Law Review* 92 (1979):547.

Breyer, Stephen, and Richard Stewart. *Administrative Law and Regulatory Policy.* Boston: Little, Brown & Co., 1979.

Brodeur, Paul. *Expendable Americans.* New York: Viking Press, 1974.

Brooks, Harvey, and Raymond Bowers. "The Assessment of Technology." *Scientific American* 22 (1970):13.

Brooks, Richard. "The Legalization of Planning." *Administrative Law Review* (1979):31.

Brown, Michael. *Laying Waste.* New York: Pantheon Books, 1980.

Buchanan, James, and W. Stubblebine. "Externality." *Economica* 29 (1963):371.

Bundy, Mary Lee, and K.R. Harmon. *The National Prison Directory.* College Park, Md.: Urban Information Interpreters, 1977.

Burger, Edward J. *Protecting the Nation's Health.* Lexington, Mass.: Lexington Books, D.C. Heath and Company, 1976.

Byse, Clark. "*Vermont Yankee* and the Evolution of Administrative Procedure: A Somewhat Different View." *Harvard Law Review* 91 (1978): 1823.

Calabresi, Guido. "Some Thoughts on Risk Distribution and the Law of Torts." *Yale Law Journal* 70 (1966):499.

Calabresi, Guido, and Phillip Bobbitt. *Tragic Choices.* New York: W.W. Norton, 1978.

Calabresi, Guido, and Jon Hishoff. "Toward a Test for Strict Liability in Torts." *Yale Law Journal* 81 (1972):1055.

Caldwell, Lynton K., Lyton R. Hayes, and Isabel M. MacWhirter, *Citizens and the Environment: Case Studies in Popular Action.* Bloomington: Indiana University Press, 1978.

California Legislature. Assembly Select Committee on the Administration of Justice. *Security and Privacy and Criminal History Information Systems.* Sacramento, March 1971.

Carroll, James D. "Participatory Democracy." In *Science, Technology and National Policy,* edited by Thomas J. Kuehn and Alan L. Porter, p. 416. Ithaca: Cornell University Press, 1981.

Carson, Rachel. *Silent Spring.* Greenwich, Conn.: Fawcett-Crest, 1962.

Catanese, Anthony, and Alan Steiss. *Systemic Planning: Theory and Application.* Lexington, Mass.: Lexington Books, D.C. Heath and Company, 1970.

Chayes, Abraham. "The Role of the Judge in Public Law Litigation." *Harvard Law Review* 89 (1976):1281.

Chen, Edwin. *PBB: An American Tragedy.* Englewood Cliffs, N.J.: Prentice-Hall, 1979.

Christensen, R."Projected Percentage of U.S. Population with Criminal Arrest and Conviction Records," In *Task Force Report: Science and Technology,* edited by the President's Commission on Law Enforcement and Administration of Justice, pp. 216–218. Washington, D.C.: Government Printing Office, 1969.

City of New York. Commission on Human Rights. *The Employment Problems of Ex-offenders.* New York, 1972.

Coase, R.H. "The Problem of Social Cost." *Journal of Law and Economics* 3 (1960):1.

Coates, Joseph F. "Technology Assessment: The Benefits . . . the Costs . . . the Consequences." *The Futurist* (December 1971):225.

———. "Technology Assessment: A Tool Kit." *Chemtech* (June 1976): 372.

———. "The Role of Formal Models in Technology Assessment." *Technological Forecasting and Social Change* 9 (1976):139.

———. "Technological Change and Future Growth: Issues and Opportunities." *Technological Forecasting and Social Change* (1977):49.

Coates, Vary T. *Technology and Public Policy: The Process of Technology Assessment in the Federal Government,* vols. 1 and 2. Washington, D.C.: George Washington University, Program of Policy Studies in Science and Technology, 1972.

———. *Technology Assessment in Federal Agencies, 1971–1976.* Report to the National Science Foundation. Washington, D.C.: National Technical Information Service, 1976.

Cobb, Roger W., and Charles D. Elder. *Participation in American Politics: The Dynamics of Agenda-Building.* Boston: Allyn and Bacon, 1972.

Cohen, Carl. *Democracy.* New York: Free Press, 1971.

Cohen, Stanley, N., et al., "Construction of Biologically Functional Bacterial Plasmids *in Vitro.*" *Proceedings of the National Academy of Sciences.* Washington, D.C.: November 1973, p. 70.

Colton, Kent W., and Kenneth L. Kraemer, eds. *Computers and Banking.* New York: Plenum Publishing Co., 1980.

Comment, "Hazardous Waste: EPA, Justice Invoke Emergency Authority, Common Law in Litigation Campaign against Dump Sites." 10 *E.L.R.* 100 (1980):34.

Conrad, J., ed. *Society, Technology and Risk Assessment.* New York: Academic Press, 1980.

Cooper, Lynne E.; M.E. Tompkins; and D.A. Marchand. *An Assessment of the Social Impacts of the National Crime Information Center and Computerized Criminal History Program,* Contractor report. Washington, D.C.: Office of Technology Assessment, U.S. Congress, October 1979.

"Costing Clones." *The Economist,* 8 August 1981.

Council on Environmental Quality. *Environmental Quality.* Washington, D.C.: Government Printing Office, 1980.

———. *Public Opinions on Environmental Issues.* Washington, D.C.: Government Printing Office, 1980.

Crandill, Robert W., and Lester B. Lave, eds. *The Scientific Basis of Health and Safety Regulation.* Washington, D.C.: The Brookings Institution, 1981.

Crenson, Matthew A. *The Un-Politics of Air Pollution.* Baltimore: Johns Hopkins University Press, 1971.

Crocker, Thomas R. "Externalities, Property Rights, and Transaction Costs: An Empirical Study." *Journal of Law and Economics* 14 (1971):451.

Culyer, A.J. *The Economics of Social Policy.* New York: Dunellen, 1973.

Daddario, Emilio Q. "Technology Assessment Legislation." *Harvard Journal on Legislation* 7 (1970):507.

Daneke, Gregory A., and Alan Steiss. "Planning and Policy Analysis for Public Administrators." In *Management Handbook for Public Administrators,* edited by John W. Sutherland. New York: Van Nostrand Reinhold, 1978.

Daneke, Gregory A., et al., eds. *Proceedings of the Conference on Public Involvement and Social Impact Assessment.* Tucson: University of Arizona, College of Business and Public Adminstration, 1981.

Davies, J.C.; S. Gusman; and F. Irvin. *Determining Unreasonable Risk under the Toxic Substances Control Act.* Washington, D.C.: Conservation Foundation, 1979.

Davis, Devra Lee; Kenneth Bridbord; and Marvin Scheiderman. "Estimating Cancer Causes: Problems in Methodology, Production, and Trends." *Banbury Report 9: Quantification of Occupational Cancer* Cold Spring, N.Y.: Cold Spring Harbor Laboratory, 1981, p. 285.

Davis, Kenneth Culp. *Administrative Law Treatise.* St Paul, Minn: West Publishing Co., 1958.

———. "Facts in Lawmaking." *Columbia Law Review* 80 (1980):931.

Davison, W.T. "Technology and Social Change." *Review of Politics* 34 (October 1972):172.

Del Duca, Patrick L. "The Clean Air Act: A Realistic Assessment of Cost-Effectiveness." *Harvard Environmental Law Review* 5 (1981):184.

Delong, James. "Informal Rulemaking and the Integration of Law and Policy." *Virginia Law Review* 65 (1979):257.

Del Sesto, Steven. *Science, Politics and Controversy: Civilian Nuclear Power in the U.S., 1946–1974.* Boulder: Westview Press, 1979.

Denison, E.F. *Accounting for Slower Economic Growth: The United States in the 1970's.* Washington, D.C.: The Brookings Institution, 1979.

———. *Effects of Selected Changes in the Institutional and Human Environment upon Output per Unit of Input.* Washington, D.C.: The Brookings Institution, 1978.

Dertouzos, Michael L., and Joel Moses, eds. *The Computer Age: A Twenty-Year View.* Cambridge, Mass.: MIT Press, 1980.

DeVine, John C. "A Progress Report: Cleaning up TMI." *IEEE Spectrum* (March 1981):44.

Dilworth, Donald C., ed. *Identification Wanted.* Gaithersburg, Md.: International Association of Chiefs of Police, 1977.

Doniger, David. "Federal Regulation of Vinyl Chloride: A Short Course in the Law and Policy of Toxic Substances Control." *Ecology Law Quarterly* 7 (1978):497.

———. *Law and Policy of Toxic Substances Control: A Case Study of Vinyl Chloride.* Baltimore: Johns Hopkins University Press, 1979.

Doub, William O. "Technological Regulation and Environmental Law." *Administrative Law Review* 26 (1974):191.

Douglas, Jack D., ed. *Freedom and Tyranny: Social Problems in a Technological Society.* New York: Alfred A. Knopf, 1970.

Dupre, Joseph, and Sanford Lakoff. *Science and the Nation, Policy and Politics.* Englewood Cliffs, N.J.: Prentice-Hall, 1962.

Edel, Abraham. *Analyzing Concepts in Social Science: Science, Ideology and Value.* New Brunswick, N.J.: Transaction Books, 1979.

Edelman, Murray. *The Symbolic Uses of Politics.* Urbana: University of Illinois Press, 1964.

Editors. "The Science Court Experiment: An Interim Report." *Science* 193 (1976):653.

Egginton, Joyce. *The Poisoning of Michigan.* New York: W.W. Norton, 1980.

Ellul, Jacques. *The Technological Society.* New York: Vintage Books, 1964.

Emery, Frederick, and Eric Trist. *Toward a Social Ecology.* New York: Plenum Press, 1975.

Environmental Defense Fund. *Troubled Waters: Toxic Chemicals in the Hudson River.* Washington, D.C. 1977.

———. *Malignant Neglect.* New York: Alfred A. Knopf, 1979.

Environmental Law Institute. *An Analysis of Past Federal Efforts to Control Toxic Substances.* Washington, D.C.: Council on Environmental Quality, 1978.

Environmental Protection Agency. Office of Research and Development. *Environmental Outlook 1980.* Washington, D.C., 1980.

Epstein, Richard. "A Theory of Strict Liability." *Journal of Legal Studies* 2 (1973):151.

———. "Products Liability: The Gathering Storm." *Regulation* (1977):15.

Epstein, Samuel S. *The Politics of Cancer.* Garden City, N.Y. Anchor Books, 1979.

Estreicher, Samuel. "Pragmatic Justice: The Contributions of Harold Leventhal to Administrative Law." *Columbia Law Review* 80 (1980):894.

Etzioni, Amitai. *The Active Society.* New York: Free Press, 1968.

Feiveson, Harold A.; Frank W. Sinden; and Robert H. Socolow, eds., *Boundaries of Analysis.* Cambridge, Mass.: Ballinger, 1975.

Ferguson, Allen R. *Attacking Regulatory Problems.* Cambridge, Mass.: Ballinger, 1981.

Ferkiss, Victor C. "Man's Tools and Man's Choices: The Confrontation of Technology and Political Science." *American Political Science Review* 67 (September 1969):973.

———. *Technological Man: The Myth and the Reality.* New York: New American Library, 1969.

Fischer, Frank. *Politics, Values, and Public Policy.* Boulder: Westview Press, 1980.

Fisher, J.F. "EFT—The Decade of the 1980's: New Concepts for the World of Banking," *The Banker's Magazine* 162 (March/April 1979):21.

Fletcher, George. "Fairness and Utility in Tort Theory." *Harvard Law Review* 85 (1972):537.

Fox, Jeffrey L. "Genetic Engineering Emerges." *Chemical and Enginering News,* 17 March 1980.

Freedman, Daryl M. "Reasonable Certainty of No Harm: Reviving the Safety Standard for Food Additives, Color Additives, and Animal Drugs." *Ecology Law Quarterly* 7 (1979):245.

Freeman, A. Myrick. *The Benefits of Air and Water Pollution Control: A Review and Synthesis of Recent Estimates.* Washington, D.C.: Council on Environmental Quality, 1979.

———. *Technology-Based Effluent Standards: The U.S. Case.* Washington, D.C.: Resources for the Future, 1980.

Freeman, A. Myrick; R.H. Haveman; and A.V. Kneese. *The Economics of Environmental Policy.* New York: John Wiley, 1973.

Friedland, Steven I. "The New Hazardous Waste Management System: Regulation of Waste or Wasted Regulation?" *Harvard Environmental Law Journal* 5 (1981):89

Friedman, John. *Retracking America: A Theory of Societal Planning.* New York: Doubleday, 1973.

Friedrich, Carl. "Political Decision-Making, Public Policy and Planning." *Canadian Public Administration Review* (1971):144.

Friendly, Henry J. "The Courts and Social Policy." *University of Miami Law Review* 33 (1978):21.

Frohlich, Norman; Joe A. Oppenheimer; and Oran R. Young. *Political Leadership and Collective Goods.* Princeton, N.J.: Princeton University Press, 1971.

Fuller, Lon. "The Forms and Limits of Adjudication." *Harvard Law Review* 92 (1978):353.

Gelpe, Marcia, and Dan A. Tarlock. "The Uses of Scientific Information in Environmental Decisionmaking." *Southern California Law Review* 48 (1974):371.

Getz, Michael, and Yuh-ching Huang. "Consumer Revealed Preference for Environmental Goods." *Review of Economics and Statistics* 60 (1978):449–458.

Goldberg, Michael, and H.I. Miller. "The Role of the Food and Drug Administration in the Regulation of the Products of Recombinant DNA Technology." *Recombinant DNA Technical Bulletin* 4 (April 1981):15.

Goodfield, June. *Playing God.* New York: Random House, 1977.

Gori, Gio Batta. "The Regulation of Carcinogenic Hazards." *Science* 208 (18 April 1980):256.

Graham, Loren. *Between Science and Values.* New York: Columbia University Press, 1981.

Green, Harold P. *The New Technological Era: A View from the Law.* Washington, D.C.: George Washington University, Program on Policy Studies, 1967.

———. "The Role of Law and Lawyers in Technology Assessment," *Atomic Energy Law Journal* 13 (1971):246.

———. "Limitations of Implementation of Technology Assessment." *Atomic Energy Law Journal* 14 (1972):59.

———. "The Risk-Benefit Calculus in Safety Determinations." *George Washington Law Review* 43 (1975):791.

———. "Cost-Risk-Benefit Assessment and the Law: Introduction and Perspective." *George Washington Law Review* 45 (1977):901.

Green, Leon. "Tort Law: Public Law in Disguise." *Texas Law Review* 38 (1959–1960):1.

Green, Mark, and Norman Waitzman. *Business War on the Law.* Washington, D.C.: Corporate Accountability Research Group, 1981.

Greene, Alan M. "Nuclear Waste Disposal: A Case for Judicial Intervention." *Wayne Law Review* 26 (1979):173.

Grezeh, Ellen E. "PBB." In *Who's Poisoning America,* edited by Ralph Nader, R. Brownstein, and J. Richards, p. 60. San Francisco: Sierra Club Books, 1981.

Grobstein, Clifford. *A Double Image of the Double Helix.* San Francisco: W.H. Freeman, 1979.

Gross, Bertram. "Planning in an Era of Social Revolution." *Public Administrative Review* 31 (1970):259.

Haberer, Joseph, ed. *Science and Technology Policy.* Lexington, Mass.: Lexington Books, D.C. Heath and Company, 1977.

Haefele, Edwin T. *Representative Government and Environmental Management.* Baltimore: Resources for the Future, 1973.

Hamilton, Robert. "Procedures for the Adoption of Rules of General Applicability." *California Law Review* 60 (1972):1276.

Handler, Philip. "The Need for a Sufficient Scientific Base for Government Regulation." *George Washington Law Review* 43 (1975):808.

Hardin, Garrett. "The Tragedy of the Commons." *Science* 162 (December 1968):1243.

Hardin, Garrett, and John Badin, eds. *Managing the Commons.* San Francisco: W.H. Freeman, 1977.

Harris, Robert H. "The Tris Ban," *Science* 197 (16 September 1977):1132.

Harvard University. *Program in Technology and Society, 1964–1972, A Final Review.* Cambridge: Harvard University Press, 1972.

Haveman, Robert, and Burton A. Weisbrod. "Defining Benefits of Public Programs: Some Guidance for Policy Analysis." *Policy Analysis* 1 (1975)169.

Hays, Steven W.; E.G. Bogan; and D.A. Marchand. *An Assessment of the Uses of Information in the National Crime Information Center and Computerized Criminal History Programs,* Contractor report. Washington, D.C. Office of Technology Assessment, U.S. Congress, October 1979.

Head, John G. *Public Goods and Public Welfare.* Durham, N.C. Duke University Press, 1974.

Heintz, S. Hershaft, and G.C. Horak. *National Damages of Air and*

Water Pollution. Washington, D.C.: Environmental Protection Agency, 1976,

Hennigan, Patrick J. "Commercialization of Recombinant DNA Research: Government-Business-University Relations." Cambridge: Harvard Business School, HBS Case Services, 1981.

———. "Regulating Bio-Medical Technology: The Case of Recombinant DNA Research." Cambridge: Harvard Business School, HBS Case Services, 1981.

Henry, Nicholas. "Bureacracy, Technology and Knowledge Management." *Public Administration Review* 35 (November/December 1975):572.

Herkett, Richard. "Environmental Alternatives and Social Goals." *The Annals* (July 1979):44.

Hetman, Francois. *Society and the Assessment of Technology.* Paris: Organization for Economic Cooperation and Development, 1973.

Hiatt, Howard; James Watson; and Jay Winston, eds. *The Origins of Human Cancer.* 3 vols. Cold Spring Harbor, N.Y.: Cold Spring Laboratory, 1977.

Hiltz, S.R., and M. Turoff. "EFT and Social Stratification in the U.S.A.: More Inequality?" *Telecommunications Policy* 2 (March 1978):221.

Hjalte, Krister. *Environmental Policy and Welfare Economics.* Cambridge: Cambridge University Press, 1977.

Hoffman, Lance J., ed. *Computers and Privacy in the Next Decade.* New York: Academic Press, 1980.

Hohenemser, C., and J. Kasperson, eds. *Risk in the Technological Society.* Boulder: Westview Press, 1981.

Hood, Christopher. *The Limits of Administration.* London: John Wiley, 1976.

Hoos, Ida R. *Systems Analysis in Public Policy.* Berkeley: University of California Press, 1972.

Horowitz, Donald. *The Courts and Social Policy.* Washington, D.C.: The Brookings Institution, 1977.

Howard, J. Woodford, Jr. *Courts of Appeals in the Federal Judicial System.* Princeton, N.J.: Princeton University Press, 1981.

Hughes, James, and Lawrence Mann. "Systems and Planning Theory." *AIP Journal* 35 (1969):330.

Humes, K.H. "The Checkless/Cashless Society? Don't Bank on It!" *The Futurist* (October 1978):301.

Hunt, James W.; J.E. Bowers; and N. Miller. *Laws, Licenses and the Offender's Right to Work.* Washington, D.C.: National Clearinghouse on Offender Employment Restrictions, 1974.

Hunter B.T. *Food Additives and Federal Policy: The Mirage of Safety.* New York: Scribner, 1975.

Hutt, Peter. "Public Policy Issues in Regulating Carcinogens in Food." *Food, Drug, and Cosmetic Law Journal* 33 (1978):541.

Jaffe, Louis. "Judicial Review: Question of Law." *Harvard Law Review* 69 (1955):239.

———. "Judicial Review: Question of Fact." *Harvard Law Review* 69 (1956):1020.

———. "The Citizen as Litigant in Public Actions: The Non-Hohfeldian or Ideological Plaintiff." *University of Pennsylvania Law Review* 116 (1968):1033.

Jaffe, Louis, and Laurence H. Tribe. *Environmental Protection.* Chicago: Bracton Press, 1971.

Jantsch, Erich. *Design for Evolution.* New York: Braziller, 1975.

Jones, Charles O. *Clean Air: The Policies and Politics of Air Pollution Control.* Pittsburgh, Pa.: University of Pittsburgh Press, 1975.

———. *An introduction to the Study of Public Policy.* No. Scituate, Mass.: Duxbury Press, 1977.

Juergensmeyer, Julian. "Control of Air Pollution through the Assertion of Private Rights." *Duke Law Journal* (1967):1126.

Kalur, Jerome S. "Will Judicial Error Allow Industrial Point Sources to Avoid BPT and Perhaps BAT Later? A Story of Good Intentions, Bad Dictum, and Ugly Consequences." *Ecology Law Quarterly* 7 (1979): 955.

Kantrowitz, Arthur, "Controlling Technology Democratically," *American Scientist* 63 (1975):505.

———. "The Science Court Experiment." *Trial* (1977):44.

Kapp, K. William. *The Social Costs of Private Enterprise.* New York: Schocken Books, 1971.

Kasper, R.G. *Technology Assessment: Understanding the Social Consequences of Technological Applications.* New York: Praeger, 1972.

Katz, Milton. "The Function of Tort Liability in Technology Assessment. *Cincinnati Law Review* 38 (1969):587.

———. "Decision-making in the Production of Power." *Scientific American* 223 (1971):191.

Kelman, Steven. *Regulating America, Regulating Sweden: A Comparative Study of Occupational Safety and Health Policy.* Cambridge, Mass.: MIT Press, 1981.

Kemeny, John G., et al., *The Presidential Commission on the Accident at Three Mile Island: The Need for Change, the Legacy of TMI.* Washington, D.C.: Government Printing Office, 1979.

King, John Leslie, and Kenneth L. Kraemer. "EFT as a Subject of Study in Technology, Society and Public Policy." *Telecommunications Policy* 2 (March 1978):13.

King, William, and David Cleland. *Strategic Planning and Policy.* New York: Van Nostrand Reinhold, 1978.

Kling, Rob. "Passing the Digital Buck: Unresolved Social and Technical Issues in Electronic Funds Transfer Systems," unpublished paper.

Irvine: University of California, Information and Computer Science Department (June 1976).

―――. "Value Conflicts and Social Choice in Electronic Funds Transfer System Developments." *Communications of the ACM* 21 (August 1978):642.

Kloman, Erasmus H. "A Mini-Symposium on Public Administration in Technology Assessment." *Public Administration Review* 35 (January/February 1975):67.

Kneese, Allen, and Charles Schultze. *Pollution, Prices and Public Policy.* Washington, D.C.: The Brookings Institution, 1975.

Kuehn, Thomas, and Alan L. Porter, eds. *Science, Technology, and National Policy.* Ithaca: Cornell University Press, 1981.

Lambright, Henry. *Governing Science and Technology* New York: Oxford University Press, 1976.

Lane, Lester. *The Strategy of Social Regulation.* Washington, D.C.: The Brookings Institution, 1981.

Lane, Lester, and E.P. Seskin. *Air Pollution and Human Health.* Baltimore: Johns Hopkins University Press, 1977.

La Pierre, D. Bruce. "Technology-Forcing and Federal Environmental Protection Statutes." *Iowa Law Review* 62 (1977):771.

La Porte, Todd R. "The Context of Technology Assessment: A Changing Perspective for Public Organization." *Public Administration Review* 31 (January/February 1971):63.

Las Alamos Scientific Laboratory. *Alternative Policy Evaluations: Four Corners Regional Study.* Washington, D.C.: National Commission on Air Quality, 1980.

Laudon, Kenneth C. *Computers and Bureaucratic Reform.* New York: Wiley, 1974.

―――. "Privacy and Federal Data Banks." *Society* (January/February 1980):52.

Lawless, Edward W. *Technology and Social Shock.* New Brunswick, N.J.: Rutgers University Press, 1977.

Lawrence, Paul, and Jay Lorsh. *Organization and Environment.* Cambridge: Harvard Business School, 1967.

Lederberg, Joshua. "The Freedoms and the Control of Sciences: Notes from the Ivory Tower." *Southern California Law Review* 45 (1972): 596–611.

Leone, R.A., and D.A. Garvin. *Regulatory Cost Analysis: An Overview.* Washington, D.C.: National Commission on Air Quality, 1981.

Lesnick, Michael, and James Crowfoot. *Bibliography for the Study of Natural Resource and Environmental Conflict Resolution.* Ann Arbor: University of Michigan, School of Natural Resources, 1981.

Leventhal, Harold. "Environmental Decision-making and the Role of the Courts." *University of Pennsylvania Law Review* 122 (1974):509.

———. "Principles of Fairness and Regulatory Urgency." *Case Western Law Review* 25 (1974):66.

Levine, Saul. "The Role of Risk Assessment in the Nuclear Regulatory Process." *Nuclear Safety* 19 (October 1978):556.

Lieberman, Jethro. *The Litigious Society.* New York: Basic Books, 1981.

Likert, Rensis. *New Patterns of Management.* New York: McGraw-Hill, 1961.

Long, Norton E. "Rigging the Market for Public Goods." In *Organizations and Clients: Essays in the Sociology of Service,* edited by W.R. Rosengren and Mark Lefton. Columbus, Ohio: Charles E. Merrill Publishing Co., 1970.

Lorange, Peter, and Richard F. Vancil. *Strategic Planning Systems.* Englewood Cliffs, N.J.: Prentice-Hall, 1977.

Loth, David, and M.L. Ernst. *The Taming of Technology.* New York: Simon & Schuster, 1972.

Lovins, Amory B. "Cost-Risk-Benefit Assessments in Energy Policy." *George Washington Law Review* 45 (1977):911.

Lowi, Theodore. "American Business, Public Policy, Case Studies and Political Theory." *World Politics* 16 (1964):677.

———. *The End of Liberalism.* New York: W.W. Norton, 1969.

Lui, B.C., and E.S. Yu. *Physical and Economic Damage Functions for Air Pollutants by Receptors.* Corvallis, Or.: Environmental Protection Agency, 1976.

Macklin, Ruth. "On the Ethics of Not Doing Scientific Research." *Hastings Center Report* (December 1977):11.

Maloney, J. "Technology Assessment in the Private Sector." Report for the National Science Foundation. Washington, D.C.: National Technical Information Service, 1979.

Maloney, M.T., and B. Yandle. *The Estimated Cost of Air Pollution Control under Various Regulatory Approaches,* unpublished paper. Clemson, S.C.: Clemson University, Department of Economics, 1979.

Mann, Dean E., ed. *Environmental Policy Implementation.* Lexington, Mass.: Lexington Books, D.C. Heath and Company, forthcoming.

Mannenbach, Stephen F. "The Decision to Choose Nuclear Power before and after Three Mile Island." *Environmental Law* 11 (1981):421.

Manning, Bayless. "Hyperplexis: Our National Disease." *Northwestern University Law Review* 71 (1977):767.

Marchand, Donald A. *The Politics of Privacy, Computers and Criminal Justice Records.* Arlington, Va.: Information Resources Press, 1980.

Marchand, Donald A., and Eva G. Bogan. *A History and Background Assessment of the National Crime Information Center and Computerized Criminal History Program,* Contractor report. Washington, D.C. Office of Technology Assessment, U.S. Congress, June 1979.

Marcuse, Herbert. *One-Dimensional Man.* Boston: Beacon Press, 1964.

Markey, Howard. "A Forum for Technocracy." *Judicature* 60 (1977):365.

Martin, David. *Three Mile Island: Prologue or Epilogue?* Cambridge, Mass.: Ballinger, 1980.

Martin, James A. "The Proposed 'Science Court'." *Michigan Law Review* 75 (1977):1058.

Martin, James G. "The Delaney Clause and Zero Risk Tolerance." *Food, Drug, and Cosmetic Law Journal* 34 (1979):43.

Martin, Jeffrey. "Procedures for Decision-Making under Conditions of Scientific Uncertainty: The Science Court Proposal." *Harvard Journal of Legislation* 16 (1979):443.

McCoy, Charles A., and John Playford, eds. *Apolitical Politics.* New York: Thomas Y. Crowell Co., 1967.

McGarity, Thomas O. "Substantive and Procedural Discretion in Administrative Resolution of Science Policy Questions: Regulating Carcinogens in EPA and OSHA. *Georgetown Law Journal* 67 (1979):729.

McGowan, Carl. "Congress, Court, and Control of Delegated Power." *Columbia Law Review* 77 (1977):1119.

McGraw-Hill Publishing Co. *Thirteenth Annual McGraw-Hill Survey of Pollution Control Expenditures.* New York, 1980.

McRae, Alexander, and Leslie Whelchel, eds. *Toxic Substances Control Sourcebook.* Germantown, Md.: Aspen Systems Corp., 1978.

Medford, Derek. *Enviromental Harassment or Technology Assessment?* New York: Elsevier Scientific Publishing Co., 1973.

Meiners, Roger. "What to Do about Hazardous Products." In *Instead of Regulation.* edited by Robert Poole, Jr., p. 285. Lexington, Mass.: Lexington Books, D.C. Heath and Company, 1982.

Mendeloff, John. *Regulating Safety.* Cambridge, Mass.: MIT Press, 1979.

———. "Does Overregulation Cause Underregulation? The Case of Toxic Substances." *Regulation* September/October 1981):47.

Merrill, Richard A. "Risk-Benefit Decision Making by the Food and Drug Administration." *George Washington Law Review* 45 (1977):994.

Merrill, Richard A., and Michael Schewel. "FDA Regulation of Environmental Contaminants of Food." *Virginia Law Review* 66 (1980):1357.

Merton, Robert K. "The Normative Structure of Science." *Journal of Legal and Political Sociology* (1942):115.

Mesthene, Emmanuel G. ed. *Technology and Social Change.* New York: Bobbs-Merrill, 1967.

———. *Technological Change.* New York: New American Library, 1970.

Michael, Donald. *On Learning to Plan and Planning to Learn.* (San Francisco: Jossey-Bass, 1973.

Miller, Herbert S. *The Closed Door: The Effect of a Criminal Record on Employment with State and Local Public Agencies.* Washington, D.C.: U.S. Department of Labor, Manpower Administration, 1972.

Miller, Herbert S., and Marietta Miller. *Guilty But Not Convicted: Effect of an Arrest Record on Employment;* Washington, D.C.: Georgetown University Law Center, 1972.

Miller, Judith. "Nuclear Safeguards Deemed Weak by U.S. Regulatory Commission." *The New York Times,* 1 December 1981.

Miller, Neal A. *A Study of the Number of Persons with Records of Arrest or Conviction in the Labor Force.* Washington, D.C.: U.S. Department of Labor, January 1979.

———. *Employer Barriers to the Employment of Persons with Records of Arrest or Conviction.* Washington, D.C.: U.S. Department of Labor, 1979.

Mills, Edwin, S. *The Economics of Environmental Quality.* New York: W.W. Norton, 1978.

Mishan, E.D. *The Costs of Economic Growth.* New York: Praeger, 1967.

———. *Cost-Benefit Analysis.* New York: Praeger, 1976.

Mitnick, Barry M. "Incentive Systems in Environmental Regulation." *Policy Studies Journal* 9 (Winter 1980):379.

———. *The Political Economy of Regulation: Creating, Designing and Removing Regulatory Forms.* New York: Columbia University Press, 1980.

Montesano, R.; H. Bartsch; and L. Tomatis, eds. *Molecular and Cellular Aspects of Carcinogen Screening Tests.* Lyons, France: International Agency for Research on Cancer, 1980.

Morgenthau, Hans J., *Scientific Man versus Power Politics.* Chicago: University of Chicago Press, 1965.

Moskow, Michael. *Strategic Planning in Business and Government.* New York: Committee on Economic Development, 1978.

Murphy, Thomas. *Science, Geopolitics and Federal Spending.* Lexington, Mass.: Lexington Books, D.C. Heath and Company, 1971.

Mushkin, Selma J. *Public Prices for Public Products.* Washington, D.C.: Urban Institute, 1972.

Muskie, Edward. "The Role of Congress in Promoting and Controlling Technology Advance." *George Washington Law Review* 36 (1968): 1133.

Nadel, Mark V. *The Politics of Consumer Protection.* New York: Bobbs-Merrill, 1971.

Nader, Ralph; Ronald Brownstein; and John Richard. *Who's Poisoning America.* San Francisco: Sierra Club Books, 1981.

Nathanson, Nathaniel. "Probing the Mind of the Administrator." *Columbia Law Review* 75 (1975):721.

National Academy of Public Administration. *Major Alternatives for Government Policies, Organizational Structures, and Actions in Civilian Nuclear Reactor Emergency Management in the United States.*

Washington, D.C. U.S. Nuclear Regulatory Commission, January 1980.

National Academy of Sciences. *Principles for Evaluating Chemicals in the Environment.* Washington, D.C., 1975.

National Advisory Commission on Criminal Justice Standards and Goals. *Report on the Criminal Justice System.* Washington, D.C.: Government Printing Office, 1973.

National Commission of Electronic Fund Transfers. *Final Report on EFT in the United States: Policy Recommendations and the Public Interest.* Washington, D.C.: Government Printing Office, 28 October 1977.

National Economic Research Associates. *The Business Roundtable Air Quality Project: Cost Effectiveness and Cost-Benefit Analysis of Air Quality Regulations.* New York: National Economic Research Associates, 1980.

National Governors' Association. *Final Report of the Emergency Preparedness Project.* Washington, D.C.: Government Printing Office, 1978.

National Research Council. *Safety of Saccharin and Sodium Saccharin in the Human Diet.* Washington, D.C.: National Academy of Sciences, 1974.

———. *Principles for Evaluating Chemicals in the Environment.* Washington, D.C.: National Academy of Sciences, 1975.

———, Institute of Medicine. *Food Safety Policy: Scientific and Societal Considerations.* Washington, D.C.: National Academy of Sciences, 1979.

Nelkin, Dorothy. "Threats and Promises: Negotiating the Control of Research." *Daedalus* (1978):107.

———. *Controversy: Politics of Technical Decisions.* Beverly Hills: Sage, 1979.

———. "Scientific Knowledge, Public Policy, and Democracy." *Knowledge* 1 (1979):106.

Neville, Robert. "Philosophic Perspectives on Freedom of Inquiry." *Southern California Law Review* 51 (1978):1128.

Nicholsen, W.J., ed. *Management of Assessed Risk for Carcinogens.* New York: New York Academy of Sciences, 1981.

"NIOSH's Proposed Program for Addressing Potential Hazards Related to Commercial Recombinant DNA Applications." *Recombinant DNA Technical Bulletin,* 4 (July 1981):66.

Niskanen, William A., Jr. *Bureaucracy and Representative Government.* Chicago: Aldine-Atherton, 1971.

———. *Bureaucracy: Servant or Master?* London: Institute of Economic Affairs, 1973.

O'Brien, David M. "The Seduction of the Judiciary: Social Science and the Courts." *Judicature* 64 (1980):8.

O'Connell, Jeffrey. *The Lawsuit Lottery.* New York: Free Press, 1979.

Oettinger, Anthony, and John C. Legates. *Domestic and International Information Resources Policy.* Cambridge: Harvard University, Program of Information Resources Policy, 1977.

Oi, Walter Y. "On the Economics of Industrial Safety." *Law and Contemporary Problems* 38 (1974):669.

———. "Safety at Any Price?" *Regulation* (November/December):16.

Olds, F.C. "Post-TMI Plant Designs." *Power Engineering* (August 1980): 54.

———. "Emergency Planning for Nuclear Plants." *Power Engineering* 85 (August 1981):50.

Olson, Mancur. *The Logic of Collective Action.* Chicago: University of Chicago Press, 1965.

Organization for Economic Cooperation and Development. *Methodological Guidelines for Social Assessment.* Paris, 1975.

Page, Joseph A., and Kathleen A. Blackburn. "Behind the Looking Glass: Administrative, Legislative, and Private Approaches to Cosmetic Safety Substantiation." *UCLA Law Review* 24 (1977):795.

Page, Joseph A., and Peter N. Munsing. "Occupational Health and the Federal Government: The Wages Are Still Better." *Law and Contemporary Problems* 38 (1974):651.

Page, Talbot. "A Generic View of Toxic Chemicals and Similar Risks." *Ecology Law Quarterly* 7 (1978):207.

Parisi, Anthony J. "The Industry of Life: The Birth of the Gene Machine." *The New York Times* 29 June 1980.

Parker, D.B. *Crime by Computer.* New York: Scribner, 1976.

Partway, P.S. "EFT Systems? No Thanks, Not Yet." *Computerworld* 12 9 (9 January 1978):14.

Pastore, S. "EFT and the Consumer." *The Banker's Magazine* 162 (March/April 1979):35.

Peat, Marwick and Mitchell and Co. "A Survey of Technology Assessment Today, A Report for the National Science Foundation." Washington, D.C.: National Technical Information Service, 1972.

———. *EFT: A Strategy Perspective.* New York, 1977.

PEDCO Environmental. *Analysis of Emission Reductions and Air Quality Changes for Alternative Development Scenarios: Ohio River Valley Regional Study.* Washington, D.C.: National Commission on Air Quality, 1980.

Pedersen, William. "Formal Records and Informal Rulemaking." *Yale Law Journal* 85 (1975):38.

Pelst, A.R., and R. Turvey. "Cost-Benefit Analysis: A Survey." *The Economics Journal* (1965):683.

Penick, James L.; Carroll W. Pursell: Morgan Sherwood; and Donald Swain, eds. *The Politics of American Science: 1939 to the Present* Chicago: Rand McNally, 1965.

Perrow, Charles. *Organizational Analysis: A Sociological View.* Belmont, Calif. Wadsworth, 1970.

Phelan, Richard. "Proof of Cancer from a Legal Viewpoint." In *Toxic Substances: Problems in Litigation,* edited by S. Birnham, p. 155. New York: Practicing Law Institute, 1981.

Pigford, Thomas H. "The Management of Nuclear Safety: A Review of TMI after Two Years." *Nuclear News* (March 1981):41.

Pigou, A.C. *The Economics of Welfare.* London: MacMillan and Co., 1932.

Porter, Alan L.; Frederick Rossini; Stanley Carpenter; A.T. Roper. *A Guidebook for Technology Assessment and Impact Analysis.* New York: Elsevier North Holland, 1980.

Portnoy, Barry M. "The Role of the Courts in Technology Assessment." *Cornell Law Review* 55 (1970):861.

Posner, Richard. "A Theory of Negligence." *Journal of Legal Studies* 1 (1971):29.

―――. "Strict Liability: A Comment." *Journal of Legal Studies* 2 (1973): 205.

Presidential Advisory Group on Anticipated Advances in Science and Technology Task Force. "The Science Court Experiment: An Interim Report." *Science* 193 (1976):653.

Press, Frank. "An Agenda for Technology and Policy." *Technology Review* 80 (1978):51.

Prest, A.R., and R. Turvey. "Cost-Benefit Analysis: A Survey." *The Economic Journal* 75 (1965):683.

Price, Don K. *Government and Science.* New York: New York University Press, 1954.

―――. *Scientific Estate.* Cambridge: Harvard University Press, 1965.

Primack, Joel, and Frank Von Hippel. *Advice and Dissent: Scientists in the Political Arena.* New York: Basic Books, 1974.

Privacy Protection Study Commission. *Personal Privacy in an Information Society.* Washington, D.C.: Government Printing Office, July 1977.

Project SEARCH. *A Model State Act for Criminal Offender Record Information.* Sacramento: California Crime Technological Research Foundation, 1971.

―――. *Model Administrative Regulations for Criminal Offender Records Information.* Sacramento: California Crime Technological Research Foundation, 1972.

Prosser, William, "Private Action for Public Nuisance." *Virginia Law Review* 52 (1966):997.

Radian Corporation. *Analysis of Innovative Policy Alternatives of Air Pollution Control Technology.* Washington, D.C.: National Commission on Air Quality, 1980.

Rasmussen, Norman I., et al. *Reactor Safety Study.* Washington, D.C.: Nuclear Regulatory Commission, 1975.

Reagan, Michael. *Science and the Federal Patron.* New York: Oxford University Press, 1969.

Redford, Emmette S. *Democracy in the Administrative State.* New York: Oxford University Press, 1969.

Regens, James L. "Energy Development, Environmental Protection and Public Policy." *American Behavioral Scientist* 22 (November/December 1978):175.

Regens, James L., and David A. Bennett. "Environmental Quality Effects of Alternative Energy Futures: A Comment on the Use of Macon-Modeling." In *Beyond the Energy Crisis,* edited by Rocco Fazzolare and Craig Smith. Oxford: Pergamon, 1981.

Regulatory Council. *Regulation of Chemical Carcinogens.* Washington, D.C., 1979.

Rettig, R. *Cancer Crusade: The Story of the National Cancer Act of 1971.* Princeton, N.J. Princeton University Press, 1977.

Rheingold, Paul. "Civil Course of Action for Lung Damage Due to Pollution of Urban Atmosphere." *Brooklyn Law Review* 33 (1966):17.

Rheingold, Paul; Norman Landau; and Michael Canavan, eds. *Toxic Torts: Tort Actions for Cancer and Lung Disease Due to Environmental Pollution.* Washington, D.C.: Association for Trial Lawyers of America, 1977.

Rhoads, Steven E. "How Much Should We Spend to Save a Life?" *The Public Interest* 51 (1978):79.

———. *Valuing Life: Public Policy Dilemmas.* Boulder: Westview Press, 1981.

Ricci, Paolo, and Lawrence Molton. "Risk and Benefit in Environmental Law." *Science* 214 (1981):1096.

Ripley, Randall B., and Grace A. Franklin. *Congress, the Bureaucracy, and Public Policy.* Homewood, Ill.: Dorsey Press, 1976.

Rivlin, Alice. *Systematic Thinking for Social Action.* Washington, D.C.: The Brookings Institution, 1971.

Rodgers, William H., Jr. "A Hard Look at *Vermont Yankee:* Environmental Law under Close Scrutiny." *Georgetown Law Journal* 67 (1979):699

———. "Benefits, Costs, and Risks: Oversight of Health and Environmental Decisionmaking." *Harvard Environmental Law Review* 4 (1980):191.

———. "Judicial Review of Risk Assessments: The Role of Decision Theory in Unscrambling the Benzene Decision." *Environmental Law* 11 (1981):301.

Rogers, Michael. *Biohazard.* New York: Alfred Knopf, 1977.

Rogovin, Michael. et al., *Three Mile Island: A Report to the Commissioners and the Public.* Washington, D.C.: Nuclear Regulatory Commission, 1980.

Rosenberg, Maurice. "Contemporary Litigation in the United States." In *Legal Institutions Today,* edited by Harry Jones, p. 152. Chicago: American Bar Association, 1977.

———. "Let's Everybody Litigate." *Texas Law Review* 50 (1972):1349.

Rossman, L.W. "Financial Industry Sees EFT Private Laws Adequate." *American Banker* 141 (28 October 1976):2.

Rourke, Francis E. *Bureaucracy, Politics and Public Policy.* Boston: Little, Brown & Co., 1969.

———, ed. *Bureaucratic Power in National Politics.* Boston: Little, Brown & Co., 1972.

Rule, James. *Private Lives and Public Surveillance.* New York: Schocken Books, 1974.

———. *Value Choices in Electronic Fund Transfer Policy.* Washington, D.C.: Executive Office of the President, Office of Telecommunications Policy, October 1975.

Rutledge, G.L., and S.L. Trevathan. "Pollution Abatement and Control Expenditures, 1972–1978." *Survey of Current Business* 60 (1980):27.

Sachs, Joel. *Environmental Law and Practice.* New York: Practicing Law Institute, 1981.

Salsburg, David, and Andrew Heath. "When Science Progresses and Bureaucracies Lag: The Case of Cancer Research." *The Public Interest* 65 (1981):30.

Sanders, Ralph. *Science and Technology: Vital National Resources.* Mt. Airy, Md.: Lomond Books, 1975.

Sayre, K., ed. *Values in the Electronics Power Industry* (Notre Dame, In.: University of Notre Dame Press, 1977.

Scalia, Antonin. "*Vermont Yankee*: The APA, the D.C. Circuit, and the Supreme Court." *Supreme Court Review* (1978):345.

Schattschneider, E.E. *The Semisovereign People.* New York: Holt, Rinehart & Winston, 1960.

Scheingold, Stuart A. *The Politics of Rights.* New Haven: Yale University Press, 1974.

Schon, Donald. "Maintaining an Adaptive National Government." In *The Future of U.S. Government: Toward the Year 2000,* edited by Harvey S. Perloff. Englewood Cliffs, N.J. Prentice-Hall, 1971.

———. *Beyond the Stable State.* New York: W.W. Norton, 1973.

Schooler, Dean, Jr. *Science, Scientists, and Public Policy.* New York: Free Press, 1971.

Schultze, Charles C. *The Public Use of Private Interest* (Washington, D.C.: The Brookings Institution, 1977.

Schwarz, Brita. "Long Range Planning in the Public Sector." *Futures* 9 (April 1977):116.

Selznick, Phillip. *TVA and the Grass Roots.* Berkeley: University of California Press, 1949.

Shapiro, David. "The Choice of Rulemaking or Adjudication in the Development of Administrative Policy." *Harvard Law Review* 78 (1965): 1601.

Shapiro, Martin. "Toward a Theory of Stare Decisis." *Journal of Legal Studies* 1 (1972):125.

Shapo, Marshall S. *A Nation of Guinea Pigs: The Unknown Risks of Chemical Technology.* New York: Free Press, 1979.

Shils, Edward, ed. *Criteria for Scientific Development: Public Policy and National Goals.* Cambridge, Mass.: MIT Press, 1968.

Silberman, Lawrence. "Will Lawyering Strangle Democratic Capitalism?" *Regulation* (1978):15.

Simmons, Robert. *Achieving Human Organization.* Malibu, Calif.: Daniel Spencer, 1981.

Sive, David. "Environmental Decisionmaking: Judicial and Political Review." *Land Use and Environmental Law Review* 10 (1979):3.

Snow, C.P. *Science and Government.* New York: New American Library, 1962.

Staaf, Robert J., and Francis X. Tannian, eds. *Externalities: Theoretical Dimensions of Political Economy.* New York: Dunellen, 1974.

Star, Ronald. "American and Japanese Controls on Polychlorinated Biphenyls (PCBs)." *Harvard Environmental Law Review* 1 (1976):561.

Steiss, Alan, and Gregory A. Daneke. *Performance Administration.* Lexington, Mass.: Lexington Books, D.C. Heath and Company, 1980.

Sterling Hobe Corp. *Survey of Methods Measuring the Economic Costs of Morbidity Associated with Air Pollution.* Washington, D.C.: National Commission on Air Quality, 1980.

Stewart, Richard. "The Reformation of American Administrative Law." *Harvard Law Review* 88 (1975):1667.

———. "*Vermont Yankee* and the Evolution of Administrative Procedure." *Harvard Law Review* 91 (1978):1804.

Stober, Gerhard, and Dieter Schumacher, *Technology Assessment and Quality of Life.* Amsterdam: Elsevier, 1973.

Sun, Marjorie. "NIH Plan Relaxes Recombinant DNA Rules." *Science* 213 (25 September 1981):1482.

Swazey, Judith R.; J.R. Sorenson; and C.B. Wong. "Risks and Benefits,

Rights and Responsibilities: A History of the Recombinant DNA Research Controversy." *Southern California Law Review* 51 (1978):1039.

Symposium. "Law and Technology." *Southern California Law Review* 48 (1974):209.

Symposium. "The Implications of Science—Technology for the Legal Process." *Denver Law Journal* 47 (1970):549.

TACA Corporation. *The Economic Impact of Regulating Air Pollution in the Ohio Valley: Ohio River Valley Regional Study.* Washington, D.C.: National Commission on Air Quality, 1980.

"Technology Assessment Seeks Role in Business." *Chemical and Engineering News,* 28 March 1977, pp. 11–13.

Teich, Albert H., ed. *Technology and Man's Future.* New York: St. Martin's Press, 1972, 1981.

"The Devils in Public Liability Laws." *Business Week,* 12 February 1979, p. 72.

"The Future of Nuclear Energy." *Physics Today* (March 1981):48.

Thompson, Frank J. *Health Policy and the Bureaucracy: Politics and Implementation.* Cambridge, Mass.: MIT Press, 1981.

Thorelli, Hans. *Strategy plus Structure = Performance.* Bloomington: Indiana University Press, 1977.

Tocqueville, Alexis de. *Democracy in America.* New York: Vintage Books, 1965.

Toffler, Alvin. *The Third Wave.* New York: William Morrow and Co., 1980.

Trauberman, Jeffrey. "Compensating Victims of Toxic Substances: Existing Federal Mechanisms." *Harvard Environmental Law Review* 5 (1981):1.

Trauberman, Jeffrey; Stuart Dunwoody; and Amy Horne. "Compensation for Toxic Substances Pollution: Michigan Case Study." *Environmental Law Reporter* 10 (1980):50021.

Tribe, Laurence. *Channeling Technology through Law.* Chicago: Bracton Press, 1973.

———. "Technology Assessment and the Fourth Discontinuity: The Limits of Instrumental Rationality." *Southern California Law Review* 46 (1973):617.

———. "Policy Science: Analysis or Ideology?" *Philosophy and Public Affairs* (1975):66.

Turvey, Ralph. "On Divergencies Between Social and Private Cost." *Economica* 30 (1963):309.

Union Carbide. *The Vital Consensus: American Attitudes on Economic Growth.* New York: Corporate Communications Department, 1980.

Urban Systems Research and Engineering. *An Institutional Assessment of the Implementation and Enforcement of the Clean Air Act: Buffalo*

Case Study. Washington, D.C.: National Commission on Air Quality, 1980.

U.S. Comptroller General. *Development of a Nationwide Criminal Data Exchange System: Need to Determine Cost and Improve Reporting.* Washington, D.C.: General Accounting Office, 1973.

——. *Development of the Computerized Criminal History Information System.* Washington, D.C.: General Accounting Office, 1974.

——. *How Criminal Justice Agencies Use Criminal History Information.* Washington, D.C.: General Accounting Office, 1974.

——. *The Nuclear Non-Proliferation Act of 1978 Should be Selectively Modified.* Washington, D.C.: Goverment Printing Office, 21 May 1981.

U.S. Congress. Commission on Federal Paperwork. *Rulemaking.* Washington, D.C.: Government Printing Office, 1977.

U.S. Congress, Senate. Committee on Environment and Public Works. *Six Case Studies of Compensation for Toxic Substances Pollution: Alabama, California, Michigan, Missouri, New Jersey, and Texas.* Report of the Congressional Research Service. 96th Cong., 2d sess., 1980.

U.S. Congress, House. Committee on Interstate and Foreign Commerce. *Report on the Toxic Substances Control Act* 94th Cong., 2d sess., 1976.

U.S. Congress. House. Committee on the Judiciary, Subcommittee no. 4. *Hearings on Security and Privacy of Criminal Arrest Records.* 92nd Cong., 2d sess., 1972.

——. Subcommittee on Civil and Constitutional Rights. *Hearings on Dissemination of Criminal Justice Information.* 93rd Cong., 1st and 2d sess., 1974.

——. *Hearings on Criminal Justice Information Control and Protection of Privacy Act.* 94th Cong., 1st. sess., 1975.

U.S. Congress. Senate. Committee on Labor and Public Welfare. Subcommittee on Health. *Examination of the NIH Guidelines Governing Recombinant DNA Research.* 94th Cong., 1st sess., 22 September 1976.

——. *Examination of the Relationship of a Free Society and its Scientific Community.* 94th Cong., 1st sess., 22 April 1975.

U.S. Congress. House. Committee on Science and Astronautics. Subcommittee on Science, Research, and Development. 90th Cong., 1st sess., Committee Print, 1967.

——. *Genetic Engineering, Human Genetics and Cell Biology: Biotechnology.* 96th Cong., 2d sess., 1980. Supp. Rept. III.

——. *Genetic Engineering, Human Genetics and Cell Biology: Recombinant Molecule Research.* 94th Cong., 2d sess., 1976, Supp. Rep. II.

——. *Hearings on Technology Assessment.* 91st Cong., 1st sess., 1969.

U.S. Congress. House. *Presidential Control of Agency Rulemaking: An Analysis of Constitutional Issues that May Be Raised by Executive Order 12291,* 97th Cong., 1st sess., 1981.

————. *Proceedings.* 90th Cong., 1st sess., 1967.

————. *Technical Information for Congress.* 92nd Cong., 1st. sess., 15 April 1971.

————. Committee on Science and Astronautics. *Technology Assessment and Choice.* 91st Cong., 1st sess., July 1969.

U.S. Congress. Office of Technology Assessment. *Hearings: Technology Assessment Activities in the Industrial, Academic, and Governmental Communities.* Washington, D.C., 1976.

————. *Cancer Testing Technology and Saccharin.* Washington, D.C.: Government Printing Office, 1977.

————. *Technology Assessment in Business and Government: Summary and Analysis.* Washington, D.C., 1977.

————. *A Preliminary Assessment of the National Crime Information Center and Computerized Criminal History Program.* Washington, D.C.: Government Printing Office, 1978.

————. *A Tehnology Assessment of the Coal Slurry Pipelines.* Washington, D.C.: Government Printing Office, 1978.

————. *Residential Energy Conservation.* Washington, D.C.: Government Printing Office, 1979.

————. *Assessment of Technologies for Determining Cancer Risks from the Environment.* Washington, D.C.: Government Printing Office, 1981.

Verkuil, Paul. "Judicial Review of Informal Rulemaking." *Virginia Law Review* 60 (1974):185.

von Hippel, Frank. "Looking Back on the Rasmussen Report." *Bulletin of the Atomic Scientists* (February 1977):42.

Waddell, T.E. *The Economic Damages of Air Pollution.* Washington, D.C.: Environmental Protection Agency, 1974.

Wade, Nicholas. *The Ultimate Experiment.* New York: Walker & Company, 1979.

————. "Gene Therapy Caught in More Entanglements." *Science* 212 (1981):24.

————. "How to Keep Your Shirt . . . If You Put It in Genes," *Science* 212 (April 1981):26.

Weinberg, Alvin M. "Social Institutions and Nuclear Energy." *Science* 177 (1972), pp. 27–34.

Weisbrod, Burton. *Public Interest Law: An Economic and Institutional Analysis.* Berkeley: University of California Press, 1978.

Weisburger, John H., and Gary M. Williams. "Carcinogen Testing: Current Problems and New Approaches." *Science* 214 (1981):401.

Welch, Bruce L. "Deception on Nuclear Power Risks: A Call for Action." *Bulletin of the Atomic Scientists* (September 1980):40.

Werner, Lettie McSpadden. *One Environment under Law.* Pacific Palisades, Calif.: Goodyear Publishing Co., 1976.

Westin, Alan, and M. Baker. *Databanks in a Free Society*. New York: Quadrangle Books, 1972.

Whitcomb, David D. *Externalities and Welfare*. New York: Columbia University Press, 1972.

White, Irvin, L., et al., *Energy from the West: Policy Analysis Report*. Washington, D.C.: Government Printing Office, 1976.

Whiteside, Thomas. *The Pendulum and the Toxic Cloud*. New Haven: Yale University Press, 1979.

Whyte, Ann, and Ian Burton. *Environmental Risk Assessment: Scope Report 15*. New York: John Wiley, 1980.

Williams, Stephen. "Hybrid Rulemaking under the Administrative Procedure Act: A Legal and Empirical Analysis." *University of Chicago Law Review* 42 (1975):401.

Wilson, David. *The National Planning Idea in U.S. Public Policy: Five Alternative Approaches*. Boulder: Westview Press, 1980.

Wilson, James Q. *Political Organizations*. New York: Basic Books, 1973.

———. *The Politics of Regulation*. New York: Basic Books, 1980.

Winner, Langdon. *Autonomous Technology: Technology Out of Control as a Theme in Political Thought*. Cambridge, Mass.: MIT Press, 1977.

Worrell, Albert C. *Unpriced Values: Decisions without Market Prices*. New York: John Wiley, 1979.

Wright, J. Skelly. "Courts and the Rulemaking Process: The Limits of Judicial Review." *Cornell Law Review* 59 (1974):375.

———. "New Judicial Requisites for Informal Rulemaking: Implications for the Environmental Impact Statement." *Administrative Law Review* 29 (1977):59.

Wright, Susan. "The Recombinant DNA Advisory Committee." *Environment* 21 (April 1979):2.

Wulff, Keith M., ed. *Regulation of Scientific Inquiry*. Boulder: Westview Press, 1979.

Yandle, Bruce. "The Emerging Market in Air Pollution Rights." *Regulation* (July/August 1978):21.

Yannacone, Victor; Bennard S. Cohen; and Steven Davison. *Environmental Rights and Remedies*. Rochester, N.Y.: Lawyers Co-operative Publishing Co., 1972.

Yellin, Joel. "Judicial Review and Nuclear Power: Assessing the Risks of Environmental Catastrophe." *George Washington Law Review* 45 (1977):969.

———. "High Technology and the Courts: Nuclear Power and the Need for Institutional Reform." *Harvard Law Review* 94 (January 1981): 515.

Zener, Robert V. *Guide to Federal Environmental Law*. New York: Practicing Law Institute, 1981.

Index

About the Contributors

Vary T. Coates is vice-president of J.F. Coates, Inc., a consulting firm specializing in technology assessment and policy analysis. She was previously associate director of policy analysis and technology assessment for Dames & Moore; associate director of the Division of Technology Assessment, Office of Environment, U.S. Department of Energy; associate director (1969–1979) of the Program of Policy Studies in Science and Technology at George Washington University; and she has taught at Carnegie Mellon University. Dr. Coates received the B.A. from Furman University and the M.A. and Ph.D. from George Washington University. She has contributed articles to various professional and academic journals.

Kent W. Colton is currently the staff director of the President's Commission on Housing, Washington, D.C. He is on leave from Brigham Young University, where he is a professor of public management and finance at the Graduate School of Management. He received the M.P.A. from the Maxwell School, Syracuse University, and the Ph.D. from the Massachusetts Institute of Technology (MIT). He served on the faculty of MIT as an assistant and associate professor of urban studies for six years and worked as the director of housing-program development at the Boston Redevelopment Authority. He was also a White House Fellow at the Treasury Department where he worked on issues of finance, housing, and financial-institution reform. His books include *Computers and Banking: Electronic Funds Transfer Systems and Public Policy,* and *Police Computer Technology.* In addition, he has authored numerous journal articles and monographs on EFT systems, financial institutions, neighborhood revitalization, and housing finance. He is also vice-president of Public Systems Evaluation, Inc., a nonprofit research corporation that focuses on the evaluation and analysis of public-sector programs and institutions.

Gregory A. Daneke is associate professor of public policy and director of the energy/environment planning and management program in the College of Business and Public Administration at the University of Arizona. He previously taught at several institutions including American University and the University of Michigan in Ann Arbor. He received the B.A. and M.A. from Brigham Young University and the Ph.D. from the University of California (Santa Barbara). He is the author, coauthor, and editor of numerous books and articles in the area of natural-resource and energy policy, including *Energy Policy and Public Administration,* (with George Lagassa);

Energy, Economics, and the Environment; and *Performance Administration* (with Alan Steiss), all published by Lexington Books. Dr. Daneke has also served as a visiting fellow and/or consultant to several local, state, and federal agencies, including the GAO and the U.S. Departments of Energy and State.

Steven W. Hays is associate professor and vice-chairman, Department of Government and International Studies, University of South Carolina. He was formerly chairman of the Department of Public Administration at California State University, Dominguez Hills. Mr. Hays is author of numerous articles and books, including *Personnel Management in the Public Sector, Public Personnel Administration: Problems and Prospects,* and *Court Reform: Ideal or Illusion?*

Kenneth L. Kraemer is director of the Public Policy Research Organization and professor of management at the University of California (Irvine). He is an internationally known researcher in the area of information systems in local government and has written numerous articles and books including *Policy Analysis in Local Government, Integrated Municipal Information Systems, Computers and Local Government, Technological Innovation in American Local Governments, The Management of Information Systems,* and *Computers and Politics.* Mr. Kraemer has also served as a consultant to the federal USAC program and as a national expert on the panel in information technology and urban management of the Organization for Economic Cooperation and Development.

National Commission on Air Quality was established by Congress under the Clean Air Act Amendments of 1977 to make an independent analysis of air-pollution control and alternative strategies for achieving the goals of the act. The thirteen-member commission included Senators Gary W. Hart and Robert T. Stafford, Representatives James T. Brayhill and John D. Dinzell, Tom McPherson, Richard E. Ayres, Tom Bradley, Annemarie F. Crocetti, Edwin D. Dodd, Jeanne Malchon, Leonard A. Schine, John J. Shechan, and Harold Tso. William H. Lewis, Jr., served as director of the commission staff.

Office of Technology Assessment is directed by John H. Gibbons. The OTA Assessment of Technologies for Determining Cancer Risks from the Environment Advisory Panel was chaired by Norton Nelson of the Department of Environmental Medicine, New York University Medical School.

James L. Regens is currently senior technical advisor to the deputy administrator of the EPA. He is on leave from the University of Georgia where he is associate professor of political science and research Fellow in the Institute of Natural Resources. Dr. Regens received the Ph.D. from the University of Oklahoma and has authored numerous articles on policy analysis and energy and environmental policy. He has also served as a visiting research fellow and/or consultant to several state and federal agencies, including the Oak Ridge National Laboratory, the GAO, the DOE, and the Southern States Energy Board.

Deborah D. Roberts is acting assistant professor in the Woodrow Wilson Department of Government and Foreign Affairs, University of Virginia. She previously served as an associate for program development for the Brookings Institution's Advanced Study Program and as research associate at the Science and Technology Policy Center, Syracuse Research Corporation. Dr. Roberts received the B.A., M.A., and Ph.D. from Syracuse University.

Mark E. Tompkins is an assistant professor in the Department of Government and International Studies, University of South Carolina. He received the Ph.D. from the University of Minnesota and has written articles and reports on public policy in state governments and the use of information in public policy and management.

David Whiteman is an assistant professor in the Department of Government and International Studies, University of South Carolina. He received the Ph.D. from the University of North Carolina.

Bonita A. Wlodkowski is an assistant professor of political science at Columbia University, where she received the Ph.D. She has served as a policy consultant in both the public and private sectors on the formulation of science and technology policy and the implementation of health policy.

About the Editors

David M. O'Brien is an assistant professor in the Woodrow Wilson Department of Government and Foreign Affairs, University of Virginia, and during 1981–1982, he was a visiting postdoctoral Fellow at the Russell Sage Foundation, New York. He received the B.A. (1973), M.A. (1974), and Ph.D. (1977) from the University of California (Santa Barbara). Prior to teaching at the University of Virginia, he taught at the University of California and the University of Puget Sound, where he also served as chairman of the Department of Politics. Mr. O'Brien is the author of *Privacy, Law, and Public Policy* (1979) and *The Public's Right to Know: The Supreme Court and the First Amendment* (1981), as well as numerous articles in law reviews and other professional journals. He is a recipient of fellowships and awards from the National Endowment for the Humanities, the Earhart Foundation, the Russell Sage Foundation, the American Philosophical Society, and Project '87, which was sponsored by the American Political Science Association and the American Historical Association.

Donald A. Marchand is associate professor in the Department of Government and International Studies and associate director of the Bureau of Governmental Research and Service at the University of South Carolina. He received the B.A. from the University of California (Berkeley) and the M.A. and Ph.D. from the University of California (Los Angeles). He is author of *The Politics of Privacy, Computers and Criminal Justice Records* (1980), and coauthor of *Information Management in Public Administration* (1982), as well as numerous articles and reports on information policy and management. Dr. Marchand has also served as a consultant and adviser to the Commission on Federal Paperwork and the OTA, as well as to the Executive Office of the President.